M.S. SWAMINATHAN
Legend in
Science and Beyond

M. S. SWAMINATHAN: THE QUEST FOR A WORLD WITHOUT HUNGER

Series Editor: M. S. Swaminathan

(M S Swaminathan Research Foundation, India)

Published:

Vol. 1:   50 Years of Green Revolution: An Anthology of Research Papers
by M. S. Swaminathan

Vol. 2:   M S Swaminathan: Legend in Science and Beyond
by P. C. Kesavan

M.S. SWAMINATHAN
Legend in
Science and Beyond

P.C. Kesavan
M S Swaminathan Research Foundation, India

World Scientific

EW JERSEY · LONDON · SINGAPORE · BEIJING · SHANGHAI · HONG KONG · TAIPEI · CHENNAI · TOKYO

Published by

World Scientific Publishing Co. Pte. Ltd.

5 Toh Tuck Link, Singapore 596224

USA office: 27 Warren Street, Suite 401-402, Hackensack, NJ 07601

UK office: 57 Shelton Street, Covent Garden, London WC2H 9HE

Library of Congress Cataloging-in-Publication Data
Names: Kesavan, P. C., author. | Swaminathan, M. S. (Monkombu Sambasivan).
   Quest for a world without hunger ; v. 2.
Title: M.S. Swaminathan : legend in science and beyond / P.C. Kesavan.
Description: New Jersey : World Scientific, 2017. |
   Series: M. S. Swaminathan: the quest for a world without hunger ; volume 2
Identifiers: LCCN 2016034786 | ISBN 9789813200098 (hc : alk. paper)
Subjects: LCSH: Agricultural innovations--India. | Agriculture--Research--India. |
   Swaminathan, M. S. (Monkombu Sambasivan)
Classification: LCC S494.5.I5 K485 2017 | DDC 338.1/60954--dc23
LC record available at https://lccn.loc.gov/2016034786

British Library Cataloguing-in-Publication Data
A catalogue record for this book is available from the British Library.

Desk Editors: Anthony Alexander/Sandhya Devi

Typeset by Stallion Press
Email: enquiries@stallionpress.com

Dedication

Dedicated with deep respect and affection to Ms. Mina Swaminathan who has in many ways contributed to bring out the best in her husband.

Preface

World over, scientists from different fields, policy makers, U.N. Organisations, media experts, as well as enlightened lay people from all walks of life will agree that one person who has been successfully thwarting the 'Malthusian scourge' in the food front for over five decades in India is Professor M.S. Swaminathan. After breaking the yield barriers with 'Green Revolution', he has introduced 'Evergreen Revolution' as a long-lasting solution to negate the Malthusian predictions. E.O. Wilson, a Harvard biologist of rare distinction has written in his epoch-making book *The Future of Life* (Vintage Books Ltd. London, 2002) that Swaminathan's Evergreen Revolution is the best option available to humankind to feed the burgeoning millions of new mouths annually and at the same time save the rest of life on the planet as well. Professor Swaminathan's magnificent contributions to science, innovative ideas, lectures, publications and field-level action plans have not only established the food security at the national level but also made it percolate down to millions of individual households. No one else has done so much to cause the transition from a 'ship-to-mouth' existence during the 1950s and 1960s to the current implementation of the world's largest social protection programme against hunger with home-grown food. Professor Swaminathan, however, did not achieve such an incredible transition with a touch of a magical wand. He is a hardcore scientist; starting from 1950, he has painstakingly built a strong foundation of basic scientific research in genetics, cytogenetics, radiation biology, induced mutagenesis on which are borne the fruits of applied research, that is, the 'Green Revolution' of the 1960s and its refined version, the 'Evergreen Revolution', and more recently the 'Farming System for Nutrition' (FSN).

This biography captures Professor Swaminathan's scientific research as opposed to the earlier ones written in a more traditional style of biographies, describing his family life, professional accomplishments in the national and international spheres. The current one, very different in content, makes in-depth analyses of his scientific publications *par excellence* in basic sciences of relevance to contemporary social need, viz., increased productivity of cereal grain crops, so that India's image as 'begging bowl' is wiped out. Immediately after ushering in the 'Green Revolution', he engaged himself in developing the concept of 'Evergreen Revolution' to fight both the famines of food and rural livelihoods and to convert the 'Biodiversity Hotspots' into 'Biodiversity Happy Spots'.

Achieving *par excellence* in basic science research during the 1950s and 1960s got him national and international recognition in many ways such as Fellowship of the Royal Society (FRS), Bhatnagar Memorial Award, Mendel Memorial Award in Brno in 1965 (to commemorate 100 years of enunciation of Mendelian Laws of Heredity) and several other Awards, but all these appear to have been overwhelmed by his subsequent stewardship of 'Green Revolution' and its transformation into an 'Evergreen Revolution'. The younger generation ought to know Professor Swaminathan's outstanding contributions to basic science. These are relevant even today, and moreover, knowledge is continuum. Several of his innovations and discoveries in basic cytogenetics, radiation biology, induced mutagenesis, etc., have put facts in the place of fallacies and clarified ambiguities. The salient features of all these important milestones have been detailed in this book. Researchers in plant breeding, sustainability science, and related subjects would find these details very helpful.

Professor Swaminathan became an acknowledged legend in biological sciences by late 1960s. Yet, he was never an 'ivory tower Professor'. His free mingling with less-gifted scientists and people from all walks of life enhanced his understanding of the pangs of hunger caused by abject poverty and especially hardships faced by women in those circumstances. This is the reason for his harnessing science and technology towards elimination of rural poverty and hunger. His dedication to this cause coupled with a combination of intellect and hard work is legendary.

There are several reasons why Professor Swaminathan is a legend, but I would like to highlight just a couple of them. Way back in 1988, when few were concerned about the climate change risks to agriculture, Professor Swaminathan cautioned about sea level rise on account of global warming, and then he suggested anticipatory research to provide *genetic shielding* of salinity tolerance to paddy, a staple crop in the coastal areas. He walked the talk, and so the M.S. Swaminathan Research Foundation (MSSRF) has genetically engineered the salt-tolerance genes from a mangrove species, *Avicennia marina* to locally grown rice varieties. A similar approach has resulted in drought-tolerant rice strains. Yet another of his successful cases of anticipatory research came to light recently. The first issue of *Rice India* (Issue 1, January–March 2016) carried a cover story on the development of *Pusa Basmati I* and acknowledged that it was Professor Swaminathan, then Director of the Indian Agricultural Research Institute, New Delhi, who initiated the research in the 1960s to develop semi-dwarf *Pusa Basmati*. It is gathered that he went on with this research despite criticisms from some quarters and finally proved to be far ahead of his times. Today, it is the first semi-dwarf, non-lodging, high-yielding and high quality *Basmati* Rice, *Pusa Basmati I* that is earning annually close to Rs. 2,000 crores in the export market.

More recently, Professor Swaminathan has added a 'nutrition value' to the 'Zero Hunger' programme launched by the UN Secretary-General Mr. Ban Ki-moon in 2012 at the RIO+20 Conference. Professor Swaminathan has developed what he calls a 'Farming System for Nutrition' (FSN) that is designed to provide agri-horticultural remedies to nutritional maladies. Yet another attractive feature of the FSN is that it can be practised in the resource-poor smallholder farms and in the process, it injects economic vibrancy to subsistence farming.

Above all these, his other innovative ideas of a *Halophyte Garden* along the sea coast and *Below Sea Level Rice Cultivation* following the Kuttanad (Kerala) traditional system reveal his personality as an intellectual who appreciates and promotes time-honoured indigenous methods and as a legend who fulfils the social contract of science. That is something done only by one who is a legend in science and beyond.

As a doctoral research scholar of the Post-graduate school of the IARI, New Delhi (1963–1967), I have used the IARI library quite extensively. It is now a matter of great pride and joy to record that the IARI library was dedicated in the name of Professor M.S. Swaminathan in April 2016. It is now 'Professor M.S. Swaminathan Library' in IARI, New Delhi.

P. C. Kesavan

About the Author

Professor Kesavan obtained his Ph.D. degree from the Post-graduate School of the Indian Agricultural Research Institute (IARI), New Delhi in 1967 for his research on 'the mechanisms of indirect actions of ionizing radiation' under the guidance of Professor M.S. Swaminathan. During 1967–1969, he taught at the University of Calgary, Alberta, and Dalhousie University Nova Scotia, Canada. He joined the School of Life Sciences (SLS), Jawaharlal Nehru University (JNU), New Delhi in 1970 as Associate Professor.

His research at the SLS primarily focussed on the elucidation of mechanisms of chemical radioprotection, especially by caffeine and several of his research publications in low dose radiobiology led to his induction to the editorial boards of several journals such as *Radiation Botany* which became *Experimental and Environmental Botany, International Journal of Radiation Biology, Journal of Radiological Protection,* etc. He is currently an Associate Editor of *Current Science* published by the Indian Academy of Science, Bengaluru.

In 1993, the Department of Atomic Energy (DAE) appointed him as the Director of the Biosciences Group of the Bhabha Atomic Research Centre (BARC), Mumbai. He was also India's delegate at the meetings of the United Nations Scientific Committee on the Effects of Ionizing Radiation (UNSCEAR) at Vienna, Austria. After the completion of his tenure at BARC, he was given the prestigious DAE Homi Bhabha Chair in Nuclear Sciences and Sustainable Rural Development, and he moved to M.S. Swaminathan Research Foundation (MSSRF) as the Executive Director of the Foundation during 1999–2003. He has published over 150 original research papers in peer-reviewed national and international journals of repute. He has co-authored, *Evergreen Revolution in Agriculture*: *Pathways to a Green Economy* jointly with Professor M.S. Swaminathan; this was published by Westville Publishing. He was also Emeritus Professor of the Indira Gandhi National Open University (IGNOU), New Delhi, and developed a few appreciation programmes in sustainable development, and a post-graduate diploma programme in sustainability science.

Contents

Acknowledgements

It is with deep sense of gratitude and pleasure that I sincerely thank Dr. S. Rajalakshmi for her painstaking efforts in going through the manuscript and doing the needful. In fact, I have no words to express my appreciation to her, especially when the time to complete volume II was short, and my wife's hospitalisation mounted enormous stress on me.

My sincere thanks are due to Ms. J.D. Sharmila for her untiring efforts in typing and getting the manuscript in proper order. She also did a marvellous job of proofreading. I also thank Ms. Y. Dilhara Begam and Mr. R. Rajamanikkam for their help in various ways especially in organising the photographs.

The stupendous work of analysing hundreds of research papers that have gone into this biography would not have been possible but for Dr. N. Parasuraman's painstaking collection of Professor Swaminathan's research papers since 1950 and maintaining them so well.

I take this opportunity to thank the following publishers for their permission to include my papers in this publication.

Euphytica
L'energia Nucleare in Agricoltura
The Nucleus
Indian Journal of Genetics
Proceedings of the XII International Congress of Genetics

Finally, I would like to compliment World Scientific for the decision to bring out the biography of Professor Swaminathan under this prestigious series.

Chapter 1

Introduction

I was pleased to accept the responsibility of writing the biography of Professor M.S. Swaminathan FRS, the first Laureate of the World Food Prize and internationally acclaimed crusader of a 'Zero Hunger' world, with humility in my heart and a feeling of inadequacy to do a befitting job. I am like a village rustic amazed at the great intellect of the village schoolmaster. The relevant part of the poem 'The Village Schoolmaster' by Oliver Goldsmith comes to my mind.

> *While words of learned length and thund'ring sound*
> *Amazed the gazing rustics rang'd around;*
> *And still they gaz'd and still the wonder grew,*
> *That one small head could carry all he knew*

However, Oliver Goldsmith would surely have known that the rustics too imbibed some of the great intellectual capacity of the village schoolmaster due to their long association with him and that indeed is also the case with me while venturing to write Swaminathan's biography. Yet, Swaminathan despite his immense knowledge and ocean of experience has maintained deep affection for the humanity and the spirit to share his achievements with all, especially the least gifted.

The uniqueness of Professor Swaminathan's approach is that he pays concurrent attention to all the dimensions of sustainable development. The important dimensions of sustainable development are ecological, social and economic. It is also necessary that sound basic research in Life Sciences is essential to integrate social and economic components for inter-disciplinary, problem-oriented approach to solving contemporary social problems, particularly poverty, hunger, malnutrition and the resultant deprivation and violence. His fight against poverty and hunger using appropriate science and technology started in the 1950s. In fact, Mahatma Gandhi visiting Kumbakonam and staying at his residence when he was only about 6 years old had left an indelible impression on him. As one of his biographers, Dil [1] writes, *Swaminathan has been greatly influenced by Gandhiji's concept of sarvodaya (advance of all at all levels) and antyodaya (starting with the poorest person in the society).* Having worked with Professor Swaminathan initially as a student in the 1960s and thereafter having greatly benefited from his knowledge and guidance for over four decades, also endorse the statement by A. Dil.

There is ample justification for so many scientists and journalists for having written the biography of Professor M.S. Swaminathan [1–8] and for still many more eminent writers and publishers aspiring to write. In as much as he has changed the image of India which at the time of her independence and until about two following decades was a *begging bowl* to *bread basket*, he is surely worth a million biographies. Who else but M.S. Swaminathan could have been ranked in 1999 by *Time Magazine* (August 23, Asia edition) as one of the 20 most influential Asians of the 20th century alongside Mahatma Gandhi and Rabindra Nath Tagore from India in recognition of his outstanding service toward a hunger-free Asia. Who else would have been chosen and honoured as the 'First World Food Prize Laureate' for his fight against hunger and rural livelihoods. Which other 'biologist' could have been elected in 2002 as the President of the Pugwash Conferences in Science and World Affairs, established by Albert Einstein and Bertrand Russell to bring peace to the world in the era of cold war and threat of nuclear weapons. Swiftly, Swaminathan, in his capacity as President of Pugwash Conferences, widened the scope of 'threat' from just nuclear weapons to hunger and pandemic human diseases such as HIV, AIDS, T.B, etc. How true it is that with the overshoot of human *ecological footprint*, and progressive degradation of the ecological foundations of sustainable agriculture coupled with social and gender inequalities, the food and nutrition insecurity at the household level of over a billion people in the world is rapidly emerging as a far greater threat than nuclear weapons of mass destruction. Experience over several decades, at least since the beginning of the globalisation with its free but *not* fair trade, shows that accelerated economic growth is unlikely to rescue the planet and humanity which are already at the crossroads. Science-based technologies over the past two centuries have, no doubt, delayed the onset of the *Malthusian scourge*, but have at the same time caused monstrous scale of degradation of the finite resources, that the very survival of humans is now in jeopardy. In his foreword to Swaminathan's book [9], Professor Jeffrey Sachs, Director, Earth Institute, Columbia University, New York writes, *the great agronomic successes since Malthus' time, including the Green Revolution itself, have come at huge and sometime irreversible environmental costs. Even with all our technological wizardy, we have not yet conquered the Malthusian Challenge since we have not yet adopted a truly sustainable method of feeding the planet.* After discussing problems and possible solutions, Professor Sachs concludes, *Swaminathan brims with ideas, prescriptions, policy plans and experiments. He knows that we can meet the great sustainability challenges ahead, but only through tremendous will, scientific knowledge, ethical commitment and openness to partnerships and cooperation. It is a tall order, but Swaminathan has proved time and again that it can be done. Now it is the reader's turn to benefit from this wisdom, experience and compassion.*

Collapse of human civilisations through growing violence would precede mass exodus and mass extinctions as is now witnessed in the world today. All these had been foreseen by Swaminathan quite a few decades ago, and he noted that hunger and social deprivation are the major causes of violence. Quenching hunger and thirst would reduce violence and also open the eyes to the splendour of pluralism in arts, music, faith, culture, etc. The need in this regard is to understand and appreciate that the direction as well as the technologies would have to be drastically changed. It is as if the world knew that a few people can turn things around. They are the ones who would avoid highly invasive and harsh technologies which do no good to environment and to substantially large proportions of the humans, and instead, innovate eco-friendly and socially inclusive technologies to reconcile conservation and development. These are the few United Nations Jacques Cousteau Chair Professors of eco-technologies of which Swaminathan is the first among the equals. His vision of a Biovillage paradigm to conserve the natural resources and make sustainable use of them to create 'on-farm' and 'non-farm' livelihoods with market linkages for income generation to reduce the rural poverty, and to increase access to food at the individual house-hold level is also the pathway to biohappiness. No wonder that his intense love for the planet Earth and its humanity was fully recognised by the International Geographical Union presenting Swaminathan with the medal *Planet and Humanity* in 2000.

Swaminathan's approach to tackling hunger problem and achieving 'Zero Hunger' planet is evolutionary in one sense. Hunger has three dimensions, viz., *caloric, protein and micronutrient.* Of these, the caloric hunger requires most immediate attention as people without any food for several days would die of starvation. Lack of protein is a form of hunger that first results in debilitation which precedes and persists for a while before death. The 'hidden hunger' caused by lack of vitamins and micro-nutrients in the diet leads to morbidity and incapacity for productive work and life. About three decades ago, at the time he was making preparations to set up the M.S. Swaminathan Research Foundation (MSSRF) in 1988, his major focus was to enhance the *access* of the rural women and men to food, which in the nature of cereal grains was available in plenty in the country. It involved creating avenues for enhancing rural livelihoods. The cause of hunger was the lack of rural livelihoods to generate income and access to food. Over the next decade, he directed the MSSRF to launch what are called the Pulse villages. Pulse crops enrich the nitrate content of the soil with the help of nitrogen-fixing bacteria in the root modules on one hand, and provide much-needed 'protein' to the people on the other. Today, the world has woken up to this concept and the year 2016 has been declared as the International Year of the Pulses by the Food and Agriculture Organization (FAO). In the current decade, with the launch of the 'Zero Hunger' programme by the UN Secretary-General Ban Ki-moon in 2012, Professor Swaminathan has come up with a highly achievable yet remark-ably simple approach to remedy the nutritional maladies with a farming system highly suitable for resource-poor small and marginal as well as smallholder family farms. He calls it the Farming System for Nutrition

(FSN) that would include cultivating agri-horticultural crops rich in specific vitamins and micronutrients in the regions where particular micronutrient deficiency causes health impairment. Put in a simple way, it is the provision of agri-horticultural remedies to specific nutritional maladies.

Today, violence is growing in the hearts of people and terrorism is driving mass exodus of refugees from Syria, Iraq, Afghanistan, etc., to Europe and other parts of the world. The consequences could lead to exacerbating the threat to the world peace, largely because of disruption in food production and access to food. Besides these *political refugees*, there are growing millions of *environmental and climate refugees;* feminisation of rural poverty is a growing social scourge. The human population explosion beyond the carrying capacity of any given region of Earth and the climate change-related sea level rise are accelerating 'mass exodus' of people from their native areas. Long before all these became a reality, Professor Swaminathan was able to comprehend the emerging scenario of challenges and threats and hence started developing science-based eco-technologies to mitigate the intensity of the strife and pangs of hunger. His setting up of MSSRF in Chennai in 1988 with the help of several prize monies he had won was towards the realisation of the principles of what he calls 'Do Ecology' and transformation of exploitative path of development into a sustainable one. His major achievement in this regard is the transformation of the unsustainable (exploitative) 'Green Revolution' into an 'Evergreen Revolution' in order to achieve productivity in perpetuity without accompanying environmental and social harm.

As of today, there are eight different biographies of Professor Swaminathan. The authors come from different parts of the world and with diverse backgrounds ranging from science to journalism [2, 4, 5]; in addition, a doctoral thesis for Ph.D. degree of the University of Madras has made an in-depth analyses of Swaminathan's life from his childhood to his present roles in many spheres, mainly as the Founder-Chairman of MSSRF. The author, Parasuraman [6], was awarded the Ph.D. degree of the University of Madras. While Dr. Parasuraman wrote his doctoral thesis in Tamil, Erdelyi [3], a Hungarian journalist, wrote the biography of Professor Swaminathan under the title, *The Man Who Harvests Sunshine —* *The Modern Gandhi: M.S. Swaminathan*, in the Hungarian language. There is a reason for Mr. Andras Erdelyi to have referred to Professor

Swaminathan as *The Man who harvests sunshine*. Mr. Erdelyi spent some time in MSSRF to collect material for the intended biography. I told Mr. Erdelyi that Professor Swaminathan had asked the architect of MSSRF building to make provision for harvesting sunshine and also rain water. The architect was apparently perplexed, and then Swaminathan explained that he meant making provisions in the top of the building for putting up solar panels for harvesting solar energy, and rooftop system for harvesting rain water and its collection in a sump. The biographer, Mr. Erdelyi, was so greatly impressed with the fact that Professor Swaminathan had thought of clean and renewable energy as well as rain water harvesting as early as 1980s when very few people subscribed to the need for clean renewable energy and rain water harvesting. This explains the reason for a rather uncommon and descriptive title of the biography. One other biography is in Marathi [8]. The biography entitled, *Life and Work of M.S. Swaminathan: Toward a Hunger-Free World* by Dil [1] is a product of extensive research and analyses of Professor Swaminathan's life and memoir. Nevertheless, all the earlier eight biographies listed in the references at the end of Chapter 12 have not comprehensively dealt with his scientific vision right from the 1950s until now for achieving a hunger-free India and 'Zero Hunger' world, while at the same time safeguarding the integrity of the ecological foundations of agriculture and harmony among humans.

All the earlier biographies [1–8] mention details of his early days in Kumbakonam (Tamil Nadu) where he was born on August 07, 1925 as the second child of Dr. M.K. Sambasivan and Parvathi Thangammal. Dr. Mankombu Krishna Sambasivan was an eminent Civil Surgeon and was also Municipal Chairman of Kumbakonam. His outstanding social contribution, apart from saving lives of patients, was the eradication of the mosquito species causing filariasis. His success was attributed to a combination of scientific techniques, social mobilisation and enhanced awareness. Such a dynamic person unfortunately died young at just 36 years of age. After the sad demise of Dr. Sambasivan, his three sons, viz., Krishnamurthy, Swaminathan and Ramdas, and his daughter Lakshmi came under the loving care of their uncle late M.K. Krishnaswamy, who for love and support, filled the void resulting from their father's untimely death. Since other biographies, particularly those by Iyer [4] and Dil [1] have extensively covered all the essential particulars of Swaminathan's

Figure. 1.1. Professor Swaminathan with his wife Mina, and his daughters, Soumya, Nitya and Madhura (from left to right).

college and university education, his marriage with Mina, (nee Mina Boothalingam) in 1955, their three daughters Soumya, Madhura and Nitya, the same are not repeated at length in this biography (Fig. 1.1).

Therefore, the present biography analyses him more as a scientist and eco-technologist for his having achieved excellence in scientific research and teaching relevant to the social and global needs of the planet Earth which is now at the crossroads.

It is a general notion that scientists with international accolades (e.g. FRS, International Awards, etc.) and contributions worthy of Nobel Prize form a group of *ivory tower Professors* — secluded from fellow scientists with lesser accomplishments and the general public. However, Professor Swaminathan has been just the opposite of an ivory tower Professor readily lending access to every student, staff, fellow scientists and all other people at all times. A term such as *people's scientist* would aptly describe him.

The other aspect of Professor Swaminathan is that he has neither been wedded to any particular technology or scientific paradigm nor divorced from any. To him, a technology is meant to solve a problem, if it can *without* causing harm to environment and health of all forms of life including humans. Most of his eight biographies [1–8] have rightly described him as an institution builder and it is equally important to emphasise that he had equipped every new institution he has built with the state-of-the-art instruments, gadgets and equipments. Yet, he had encouraged and nourished the development of ideas and mental capacity to realise the intended goal *even without* the sophisticated tools and instruments. He was so right when Bhabha Atomic Research Centre (BARC), Mumbai developed 'parallel processing' to enhance speed and accuracy of computations, when super computers were denied by the super powers. Self-reliance is the key to sustainable development.

Planet Earth was already at the crossroads when he set up inter- and trans-disciplinary research programmes at MSSRF in 1988. As far as the science-based technologies for the rural areas are concerned, he blended the frontier technologies with traditional knowledge and ecological prudence of the rural and tribal people, particularly women. The resultant eco-technologies with their *pro-nature, pro-poor, pro-women* and *pro-livelihood* orientation fulfil the ecological, economic and social needs of any sustainable development. While this motto of Professor Swaminathan is for technology development towards sustainable management of natural resources in the rural areas and creation of *on-farm* and *non-farm* livelihoods, changing the *direction* of scientific queries to provide solutions to the existing as well as emerging problems of the resource-poor small and marginal farms and farmers was also integrated in his programmes of *social contract of science and technology.* This reminds us of the old Chinese proverb that *if we do not change our direction, we are likely to end up where we are headed.*

His goal throughout the past 65 years of his journey in science has been to achieve a world that would not have any one of the three kinds of hunger viz., *caloric, protein* and *micronutrient.* There is no magic wand to achieve these; on the other hand, the use of technologies which are inherently invasive and harmful to the ecological foundations of sustainable agriculture needs to be prudently avoided or limited to situations of extreme necessities. Discretion, therefore, to weed out undesirable technologies, and promote only such technologies which would strengthen/promote sustainable agriculture and rural development is highly essential. Unless this exercise is done before harnessing a technology, irreparable damage to environment and 'life' on Earth would be inevitable. It is indeed unique that Professor Swaminathan's transformation of the 'Green Revolution' into 'Evergreen Revolution', and his concepts of 'Halophyte Garden', 'Rice Cultivation Below Sea Level', etc., provide 'doable' pathways to achieve sustainable agriculture and rural development even in an era of 'global change', of which climate change is an important integral. These are just a few examples of immensely vast approach to achieving 'Do Ecology' of which he is the architect.

Ideas and adoption of simple tools have played a significant role in the evolution of humans from prehistoric times to the 21st century. Simple techniques often play major roles in changing human destiny. For example, I remember Professor Swaminathan once telling that Sir C.V. Raman used simple tools to discover the 'Raman Effect' which won him the Nobel Prize. It is known that he used equipment hardly worth Rs. 200 to make his epoch-making discovery. In his own research as a young scientist, Swaminathan has used very simple techniques to overcome major barriers to inter-specific crossability in *Solanum* section *Tuberarium*. His accomplishments in basic research in cytogenetics, induced mutagenesis with ionizing radiation and chemical mutagens, organisation of chromosomes have been highly impressive and remain relevant even after six decades. A brief account of these was brought out in the section on 'Living Legends in Indian Science' — a series published by *Current Science*. Writing under the title, *M.S. Swaminathan: A journey from the frontiers of life sciences to a state of a 'Zero Hunger' world*, Kesavan and Iyer [10] have highlighted some of his outstanding basic research contributions, these in turn opening up new avenues for applied research in order to enhance agricultural productivity, and build human resource base capable of meeting the anticipated problems in the near and distant future.

In comprehending the problems yet to reach our doorsteps, Professor Swaminathan has always been quite ahead of the rest. Not only did he clearly foresee the adverse impact of anthropogenic climate change on agriculture way back in the 1970s and 1980s, he had also suggested different science-based strategies as well as village level codes in order to reduce the yield losses during bad monsoons, and maximise the gains during good monsoons; his idea of Gene-Seed-Grain-Fodder-Freshwater banks (banks with a difference) elegantly integrates conservation and food security at rural household levels, particularly during extreme natural events both geophysical (e.g. earthquakes) and hydrometeorological (i.e. floods, drought, etc.).

There is a saying as follows:

People who change AFTER change — SURVIVE
People who change WITH change — SUCCEED
People who cause the change — LEAD

It goes without saying that those who lead are often the target of undue criticism and vicious attacks in media and elsewhere. Swaminathan had always been subject to uncharitable witch-hunting, particularly whenever he was to be elevated professionally or chosen for a prestigious award. Love is the weapon that Swaminathan uses against those who exhibit intolerance, use harsh words and sabotage his progress in work. Although Swaminathan rose rather rapidly in ranks and with each rise, he shouldered enhanced administrative responsibilities, he has remarkably balanced his authority with compassion. He adhered to the rules and policies, yet he also saw *human face* and showed lenience wherever justified. I gather that on one occasion, one of his senior scientists overstayed abroad for a day after attending an international scientific conference. As the Head of the Division, Swaminathan faced silly questions from bureaucracy in higher levels as to the actions proposed to be taken against the erring scientist. As the written note filings built up, one final question to Swaminathan was *whether the overstay for a day by the scientist concerned was inevitable.* Swaminathan in a terse but polite manner wrote back, *nothing but death in life is inevitable.* While this case is only illustrative, there have been many other instances wherein he went into the root cause of an incident or situation that necessitated a deviation from the rule book.

This introduction, however long, would still be incomplete if it does not make a reference to the cordial and, in fact, affectionate relations brought about by Mrs. Mina Swaminathan between her family and the large number of research scholars, students and colleagues of her husband who had already been widely known globally for his outstanding scientific contributions and greatly respected. During my days at the Post-graduate School, Indian Agricultural Research Institute (IARI), the Bungalow No. B-12, the official residence of Swaminathan was an epicentre of several creative yet relaxing activities. The research scholars and close associates of Swaminathan would be invited to the Birthday and New Year parties organised by Mina. Lots of games and delicious food items spread over tables were always the major attraction to students. On the last day of the calendar year (i.e. December 31) an event called Botany Division in Retrospect would be convened to take stock of scientific achievements as well as setbacks and challenges. It would end with a Cultural Programme in the evening organised by the students and staff of the Botany Division. Mina's contribution from behind the curtain would invariably be the secret of the grand success of the programme.

B-12 was always serene for its harmony with nature and humanity. The living room had simple but elegant furniture and chosen paintings which reflected the intellectually oriented way of life of Mina and Swaminathan. The piano in the living room gave the feeling of it playing Mozart even without Mina operating the keys. Mina had organised a Music Club and top musicians of the day were periodically invited to perform. A wide spectrum of Carnatic, Hindustani, Classical and also light music by young non-professional music lovers brought out the rich diversity of music in the country.

Mention has been made by Iyer [4] about Mina having been student of M.A (Economics) at Newnham College, University of Cambridge, UK during 1951–1953 when Swaminathan as a Ph.D. scholar at the School of Agriculture was also at the same university. They had met on a few occasions but did not decide to marry then. About 2 years later, back in India, they were married in 1955. Years later, Swaminathan reminisced that the Cambridge acquaintance was a prime factor in their later decision to marry.

Mina, as I knew since 1963, is a great teacher and powerful speaker. She is known for her command over both spoken and written English. During my student days at the IARI, I have seen some of her students freely interacting with her, and spending time at her residence.

Mina had distinguished herself as an adorable and inspiring teacher in her early life, and she evolved over the years as an eminent authority in the field of education with particular reference to preschool education. She played a key role in the development of Integrated Child Development Services (ICDS) as well as Mobile Creches catering to the needs of migrant labour children.

Mina is a voracious reader, especially the science fiction books. She is adept in using formal science wherever necessary. In more recent times, she has been devoting a great deal of her time to fight for decriminalisation of homosexuality, particularly in the Indian context. Her quest was whether homosexuality is gene-based, and if so, how it is inherited. Search led her to the prevalent knowledge that the trait of homosexuality is not genetic but *epigenetic*. What this means is that there are no identifiable genes as such, but the higher levels (concentration) of testosterone imprinting on the developing foetus confers inclination toward same sex in the adolescent and adulthood life. She is right that homosexuals are physiologically made that way to prefer same sex mates in life. Homosexuality therefore is not a criminal act.

Mina's research and development work since long has been focused on gender equality, rights of women farmers (who have legally no right to land ownership), feminisation of rural poverty, and voicing the voiceless women who are subjugated in one way or another. No one else has done more to articulate the invisibility of women's work than Mina. In fact, she has gone on to put on record that housewives, who are often regarded to be sitting idle and not working in the office as their husbands, actually do lot more work than their spouses. Her vigorous pursuit has opened the eyes of many men to the reality of the amount of work their wives do in the house. That is indeed social reformation in the contemporary times.

Both Mina and Swaminathan walk the talk. Their daughters were free to choose their husbands. In traditional South Indian families, the possible roles and interplay of stars, planets and moon are first ascertained through an astrologer before alliances for marriage are ventured. They instead followed the dictum that *marriages are made in heaven.*

Swaminathan on numerous other occasions have lived up to Mahatma Gandhi's words *Be the change you want.* It is the freedom and way of life at home that have made their three daughters most outstanding in their own professions and cordial as well as caring in the social relations. The eldest daughter Soumya is now the Director-General of the Indian Council of Medical Research — she is the second woman DG-ICMR. The second, Madhura started her distinguished career in economics as a Rhodes Scholar, now Professor at the Indian Statistical Institute, Bengaluru and is also the Chairperson of MSSRF. The third and the youngest daughter Nitya is Professor of Gender and Development at the University of East Anglia, Norwich, UK.

Mina, Swaminathan and their three daughters are also lovers and promoters of music. Madhura used to perform Odissi dance. Currently, Swaminathan avails every opportunity to attend Shri T.M. Krishna's music concert. Mention was made earlier to the piano in the living room of B-12 residence of Swaminathan at IARI, New Delhi. To me, it also symbolised keeping the window open to let other cultures and music in, as once said by Mahatma Gandhi.

Two major events took place in IARI in 1965 that needs to be recorded. One was the International Award namely *Mendel Centenary Medal of Czechoslovak Academy of Sciences* to Swaminathan for his outstanding contributions to Plant Genetics; the other was the ballet on DNA double helix developed and directed by Mina. Rendering the structure of the double-helix DNA and its replication before cell division in the form of a ballet was absolutely ingenious and has no parallel anywhere to the best of my knowledge. Placing young men and women as the purine and pyrimidine bases, their forming the double helix with hydrogen bonding, and their separation (i.e. unwinding) to make room for incorporation of newly synthesised bases in a semi-conservative manner, and finally the reformation of two DNA helices in the place of one original helix seem all far too complicated to me even today. It is truly difficult to comprehend how she did what she so elegantly did. No doubt, many foreign geneticists who participated in the Genetics symposium were greatly amazed and wonder struck. Ralph Singleton, a US geneticist made a note of it in his textbook *Elementary Genetics* published by D.Van Nostrand Company Inc., 1967.

References have already been made to the eight biographies of Swaminathan. Subsequent to these, a lot more information has been brought to light in the articles written by some of the world's most distinguished scientists [11]. There have been several previously unknown/not well-known and highly interesting anecdotes and endearing details in some of those articles. For instance, Ambassador Kenneth M. Quinn, President, World Food Prize, Des Moines, USA has begun his article with a photograph of a large mural of Swaminathan in the World Food Prize Hall of Laureates. This mural in pleasing colour combination is absolutely adorable.

Yet another by Professor Louise O. Fresco, Vice Chancellor of Wageningen UR, and the Netherlands has made quite a touching beginning [12], as quoted below:

It is the year 1949. The Netherlands is still recovering from the German occupation. At the small station of Ede-Wageningen a rather un-Dutch looking slender young man steps down from the train, a heavy suitcase in his hand. He looks around to find his host, but sees no one. Only a tall man who, with a firm gesture takes his suitcase from him. A porter, the young man thinks and feels embarrassed because he has no Dutch money on him as he came by overnight ferry from England and stepped immediately on the train upon arrival in Hoek van Holland. How is he going to tip the porter? Putting aside his diffidence, he asks the man if he perhaps knows how to get to Professor Prakke at the Agricultural University in Wageningen. I am Professor Prakke, the man replies, so the embarrassment of the young man turns into astonishment. A professor who picks up a young researcher from the station and starts carrying his suitcase is something he could have never imagined (R. Rabbinge, pers. commun.).

It is in the knowledge of most people that Swaminathan and Borlaug worked together for ushering in India's Green Revolution of the 1960s, but it is not widely known that Swaminathan referred to it as 'exploitative agriculture' and that if it is practised for long period of time without adherence to scientific principles, it would result in agricultural doom than prosperity. In fact, that was the theme of his address in January 1968 at the *55th session of the Indian Science Congress* held in Varanasi. He was also concerned that cultivation of a few high yielding varieties contiguously over large areas could enhance susceptibility to pests and diseases which in turn could cause food famines as it happened in Ireland (Irish potato famine 1845–1849) and in India (Bengal rice famine of 1943–1944). In addition, giving up cultivation of the numerous locally adapted varieties and land races with preference only to a few high yielding varieties would result in loss of agro-biodiversity. As had been predicted by him, it soon became evident that soils, biodiversity and freshwater sources (ecological foundations of agriculture) have been progressively undergoing degradation and in the 1990s, papers were published confirming that yield gains associated with Green Revolution have already begun to show signs of fatigue. During the 1980s, it also became evident that 'Green Revolution' (a term coined by Dr. William Gaud of the US Agency for International Development in 1968) had, no doubt, established food security at the national level but failed to do so at the individual household levels of hundreds of millions of rural people. The paradox *Mountains of grains on one hand and millions of hungry on the other* aptly described the scenario.

It had been known for quite some time that hunger in India was largely due to lack of *access* (i.e. lack of purchasing power) to food than non-availability of food grains. This observation necessitated an approach that the science and technologies developed for India's rural areas should facilitate solving the famines of food and rural livelihoods concurrently. Having taken all these into account, Swaminathan developed the concept of 'Evergreen Revolution' and defined it as *achieving productivity in perpetuity without accompanying ecological and social harm.* Among the first to applaud Swaminathan for his eco-friendly agriculture and sustainable rural development strategy was Professor E.O. Wilson of Harvard University. In his epoch-making book *The Future of Life* [13] he has emphasised that Swaminathan's 'Evergreen Revolution' is the best option available to humankind to feed the burgeoning millions new mouths and also at the same time, save the rest of life without entering into a Faustian bargain.

Translation of new knowledge generated in the laboratory research into yield gains in the farmers' fields, especially of the resource-poor small and marginal farmers requires appropriate institutions with lab to lab, lab to land, and land to lab linkages. In this regard, Swaminathan is a great institution builder. The nature of institutions he has built reveals him also as a social scientist *par excellence*.

This biography has 12 chapters. These are set in a manner that reflects his own evolution or rather transition from basic research to intensely applied research inclusive of the principles, tools and techniques of the social sciences. Gender concerns are integral to all his endeavours. There has been discernible divergence over the years in approaches towards the goal of food security by Borlaug and Swaminathan. This needs to be read in the true spirit of adherence to environmental, social and economic dimensions of sustainable development. In fact, Professor Rudy Rabbinge writes, *The role of breeding as the engine behind the Green Revolution is in many cases overestimated, but not by Swaminathan. He made it that the combination of disciplines was needed and laid the basis for the substantial increase of crop productivity* [14]. Professor Rabbinge's article was aptly titled as *M.S. Swaminathan: His Contributions to Science and Public Policy*. It could be that Borlaug was in a desperate haste to produce more food while paying little attention to environmental and social detriments, whereas Swaminathan had included all the three major pillars of sustainable agricultural productivity. Dennis Garrity in a personal communication (2016) has adopted Swaminathan's approach (called Evergreen Revolution) to change the agricultural destiny of Africa for better. He exclaims *Evergreen Revolution truly works* and is developing a large programme that plans to reach 50 million African farmers immediately and 100 million hectares of land by 2030. In a recent email to Dr. Garrity, Swaminathan states, *The concept of Evergreen Revolution is consistent with the UN Sustainable Development Goals.*

As the several chapters of the book would show, Swaminathan's initial tryst with cytogenetics of potatoes has taken him to unprecedented new heights of developing strategies to achieve 'Zero Hunger' in a world in which human population is progressing geometrically, the finite resources for production of human needs, particularly the food, are rapidly diminishing, while the threat of arriving at the 'tipping point' due to

Anthropogenic climate change is real. As an editorial in *Nature* put it about half a decade ago, the climate science and climate politics seem to be contained in two parallel worlds. The point of concern is that Albert Schweitzer's statement appears almost inevitable. He said: *Man has lost the capacity to foresee and to forestall. He will end by destroying the Earth.* Sir Martin Rees of the Institute of Astrophysics at the University of Cambridge, UK also arrived at the same conclusion of inevitable doom in his book *Our Final Hour* [15]. Yet, Swaminathan believes that the humanity and the planet Earth could still be saved by adopting a way of life in harmony with nature and embarking on a path of sustainable development. Finally, it must be noted that as a citizen of a democratic nation, he could only show what is good and doable, and it is left to the elected governments to accept or not the wisdom and experience of Professor Swaminathan. For instance, the number of farmers' suicides in India would have been drastically reduced, had his report as the Chairman of the National Commission of Farmers (NCF) been implemented soon after it was submitted in 2004. The NCF under Swaminathan's Chairmanship had recommended in 2006, the initiation of a conservation farming movement in the heartland of the Green Revolution in order to halt the damage to the ecological foundations of sustainable agriculture.

In the more recent times, we read about the farmers' suicide almost every day. Land is rapidly becoming a shrinking resource and availability of freshwater to agriculture is diminishing day by day. National resources are not being managed in a manner for future generations to meet their own needs of food, fibre, shelter, etc. No wonder Swaminathan had unambiguously expressed himself many a time that *if agriculture goes wrong, nothing else will go right.* Hopefully, this book, yet another biography, might reignite the minds of the youth to save the planet, humanity and all other forms of life. Food is and will always be the most essential basic and inevitable need among the hierarchy of human needs. Swaminathan's journey in science for over 65 years has been to achieve productivity in perpetuity and enlarge the resource base for on-farm and non-farm rural livelihoods on the one hand, and help the national governments to frame policies towards hunger-free nations and a 'Zero Hunger' world on the other.

Chapter 2

Contributions to Cytogenetics of Cultivated Potato and Its Wild Relatives

2.1 Potato and Hunger of the Poor

It is an uncanny coincidence that Swaminathan, a Gandhian at heart with utmost concern for the poorest and the weakest, started his research career over 65 years ago in potato which is the staple food of the poor across the western world in general and the people of the poorest country in Western Europe, namely Ireland in particular, during the 16th to 19th centuries. For instance, The French sociologist, Gustave de Beaumont, who visited Ireland in 1835 [16] wrote: *I have seen the Indian in his forests, and the Negro in his chains, and thought, as I contemplated their pitiable condition, that I saw the very extreme of human wretchedness; but I did not then know the condition of unfortunate Ireland ... In all countries, more or less, paupers may be discovered; but an entire nation of paupers is what was never seen until it was shown in Ireland.* So, the potato nicknamed as 'the root' keeps together the body and the soul of the very poor people.

Potatoes are not native to Ireland but probably originated in the Lake Titicaca region on the Peru–Bolivian border. Spanish conquerors found the Incas growing the tubers, which the Spanish called *patata*. They were taken back to Europe and eventually reached England where the name changed to *potato*. About 1590, potatoes were introduced to Ireland where farmers quickly discovered that potatoes thrived in their country's cool moist soil with very little labour. An acre of fertilised potato field could yield up to 12 tonnes of potatoes, enough to feed a family of six for a year with leftovers going to the family's animals.

By the early 1800s, potato had become the staple crop in the poorest regions. More than three million Irish peasants subsisted solely on the vegetable which is rich in protein, carbohydrates, minerals, and vitamins such as riboflavin, niacin and vitamin C. It is possible to stay healthy on a diet of potatoes alone. The Irish often drank a little buttermilk with their meal and sometimes used salt, cabbage, and fish as seasoning. Irish peasants were actually healthier than peasants in England or Europe where bread, far less nutritious, was the staple food.

The book, *The History and Social Influence of the Potato* by Salaman [16], is quite comprehensive especially in its analysis of the importance of potatoes in food and culture of people all the way from the Andean highlands to its tragic role in Ireland. Potato has also been eulogised by Vincent Van Gogh through his painting, *The Potato Eaters* in 1885. Potato was not only the staple food of the poor but also demeaned as 'just root' by the wealthier English in the 17th century. An interesting anecdote narrated by Salaman in his book [16] explains how potato was assigned possibly the lowest level in the hierarchy of human food basket by the English. It is the story of an Englishman ship-wrecked in Ireland who took shelter in a parson's house near the church. The parson offered him food whatever he could spare. Later, back in England, the Englishman recorded the hospitality of the parson and referred to 'roots' having been served along with fish for dinner. Years later, it was clarified that the term 'roots' actually referred to potatoes.

Salaman also describes the depressing socio-economic conditions in Ireland in the middle of the 16th century when potato was introduced into the country. He describes in the following words: The *people were famished; to sow their usual crops, was but to invite their destruction. Every seed crop, be it oats or barley, rye or wheat, might be trampled over and ruined in a day; if it escaped that hazard, the garnered harvest might be raided or burnt overnight. The vegetable crops, cabbage and parsnip, were no less vulnerable, at best they were but auxiliary foods, and there was never much of either. It was under such conditions that the potato made its entry into Ireland. Its greedy acceptance by the people was no mere accident, for it satisfied their needs as efficiently as it symbolized their helpless degradation.*

Several biographers have observed that Swaminathan's decision to study agriculture and not medicine was to a great extent influenced by the Bengal famine in 1944–1945. He with a Gandhian mindset would have been greatly agonised by what he had surely read about the Irish potato famine of 1845–1849 that occurred about 100 years earlier. So, it is inferred that when Swaminathan received UNESCO, The Netherlands Government Fellowship in 1949 to study at the Wageningen University, he possibly accepted it guessing that he might have to work on potato. His approach was to use the best of science of his time during the 1950s and do whatever feasible to avert the recurrence of yet another potato famine as intense as that of the 1845 or even worse.

Swaminathan possibly had also thought of the adverse impact of the potato famine. During the period of famine (1845–1849), about one million Irish people died of starvation and about an equal number of them crossed the Atlantic to the New World. Many of the descendants of the migrants to the North America, *the land of opportunities*, greatly flourished and form the top echelon of the present Great American Society. It is rather ironical that the Irish potato famine of the 19th century was in a way responsible for the descendants of the migrants to achieve fame, wealth and influential power in the New World. The genetic and cytogenetic research in potato and its wild relatives in the USA may at least in part have been triggered by a sense of gratitude beyond the fact that potato in the American meal was very common and a source of starch.

There is a divergence between notion and science as to when one played a major role in the Irish potato famine of 1845. The scientific reason of the famine was the *late blight of potato* caused by fungi called *Phytophthora infestans*. The other reason, mostly socio-economic, advanced by the English reformers is that Ireland's 'surplus' population doubled to over eight million before the famine. Even with eight million people, the notable cause of starvation even 'before' the late blight attack was the lack of livelihoods and 'access' to food. The British 'Poor Enquiry' survey, conducted in 1835, revealed that 75% of Irish labourers were 'without' any regular work and that begging was very common. Further, it must also be emphasised that there had been a few potato famines before that of

1844–1845. November of 1739 witnessed one of the severest frosts that Ireland had ever experienced. It destroyed the potatoes both in field and clamp and produced widespread famine; a conservative estimate was that about 3,00,000 people perished in hunger. Frost induces what is referred to as 'abiotic stress' and the pathogenic fungus (e.g. *P. infestans*) induces 'biotic stress'. Nematodes (cylindrical worms which cause damage to plants) also induce 'biotic' stress. Swaminathan's tryst with cytogenetics of potato and its wild species (1950–1954) led to the development of 'genetic shield' against both the biotic and abiotic stresses which is discussed later.

Although the Irish potato famine that occurred during 1844–1849 could certainly not have influenced Thomas Malthus's *An Essay on the Principle of Population* (1798), it possibly induced Malthus to write, *Principles of Political Economy* in 1820. Salaman considers that Malthus may well have had the condition of Ireland in his mind when he wrote:

> *If every person were satisfied with the simplest food, the poorest clothing and meanest houses, it is certain that no other sort of food, clothing and lodging would be in existence, and as there would be no adequate motive to the proprietors of land to cultivate well, not only wealth derived from convenience and luxuries would be quite at an end, but if the same divisions of land continued, the production of food would be prematurely checked and population would come to a stand long before soil had been well cultivated.*

Professor Swaminathan's entire research in the cytogenetics of potato was to a great extent moulded and directed by the poverty, hunger and deprivation of an unprivileged section of humanity in the Europe, details of which were drawn from his reading of Salaman's *The History and Social Influence of the Potato* [16]. Yet another reason possibly was that potato is nutritionally superior to oats, wheat and barley. Further by 1950, the demand for potato world over had grown tremendously. Value-added products (e.g. chips) came to be greatly sought after as a snack. In the UK, fish and potato chips are favourite fast food. During summer months in Nova Scotia, Canada, having 'fish and chips' as lunch on the beach remains fresh in my memory. Children love potato chips and these are relatively affordable.

In a lighter vein, one could say that more than The Netherlands Government, it is the potato that beckoned Swaminathan to come to Wageningen and improve its *genetic* lot. It is imagined that potato might have told Swaminathan the following: *I have done more than enough to keep the body and soul of the poor than wheat and oats. Yet, the new cadre of scientists calling themselves as geneticists and cytogeneticists are carrying out several sorts of cytogenetic research studies on wheat, oats and barley both in the labs and on field. As you know, these crops stand erect boasting their panicles with grains whereas I remain unseen covered by soil under the ground. Yet another reason is that most of these scientists are unfamiliar with genetic improvement of vegetatively propagated crops. I know you are quite different and will accept all the challenges that my inherent nature could pose to you. My future generations will owe it to you.*
Thus, Swaminathan landed in Wageningen in 1949.

2.2 An Introduction to Polyploidy Since Potato is a Tetraploid

Within 6 months of his arrival in Wageningen, Swaminathan authored his first scientific article in German language [17]. This paper elegantly describes the simple technology to stain and spread the potato chromosomes with great clarity so that their numbers and structures could be accurately assessed under the light microscope. His cytogenetic investigations involved not only the cultivated potato, a natural tetraploid ($2n = 48$), but also its wild relatives consisting of hexaploids ($2n = 72$), tetraploids ($2n = 48$), and diploids ($2n = 24$). The term 'diploid' denotes the organisms with only two sets of chromosomes, of which one set is derived from the male and another from the female parent. The diploid numbers of chromosomes widely vary among the different genera and species. A complete set of chromosomes in a diploid is called a genome. Few examples to illustrate this are: The cultivated barley (*Hordeum vulgare*) has chromosome complement of $2n = 14$. Of these, a set of seven chromosomes has come from its female parent and the other set of seven chromosomes from its male parent. Thus, it has a 'diploid' complement ($2n$) of 14 chromosomes; the complement of just one set or the 'haploid' complement (n) is of seven chromosomes. Onion, *Allium cepa* has diploid number of 16 chromosomes ($2n = 16$). Cultivated cotton is of interest. It has a few species such as *Gossypium arboreum* with a somatic complement of 26 chromosomes ($2n = 26$). This species is native to India. On the other hand, the New World Cotton, *Gossypium hirsutum* has a somatic complement of $2n = 52$ chromosomes. There are several agriculturally important crop plants which have somatic complement of chromosomes that are multiples of their diploid relatives. Cultivated wheats have $2n = 28$ and $2n = 42$, whereas their wild progenitors have $2n = 14$ chromosomes. With wide occurrence of polyploids along with diploids among the flowering species, the term haploid complement of chromosomes ('n') becomes prone to confusion and misunderstanding. For example, how do we say that the haploid complement (n) for *G. arboreum* is 13, and that of *G. hirsutum* is 26 without wondering about the presence of two different haploid complement numbers for the genus *Gossypium*. In such situations, it would be erroneous to say that the haploid (n) number for the genus *Gossypium* is 13 or 26. The cytogeneticists have, however, resolved the problem by introducing the term 'basic' number in addition to the 'haploid' number. The 'basic' number of chromosomes of the species is denoted by 'x'. With this refinement, the genomic constitution of the diploid cotton *G. arboreum* is written as $2n = 2x = 26$ and that of the tetraploid *G. hirsutum* as $2n = 4x = 52$. The haploid of a tetraploid is actually composed of two basic sets of chromosomes ($2x$). It also reveals that the haploid number (n) of a tetraploid cotton as 26 is a pseudo-haploid. Commonly, a better term *polyhaploidy* is used to denote the haploid obtained from polyploid species. Conservation of the diploid and polyploid wild relative species in their primary and secondary Centres of Origin (i.e. *in situ* conservation) has been one among several major crusades of Swaminathan. He has often pleaded saying that these wild related species contain genes to shield against a variety of abiotic (e.g. salinity, submergence, drought, etc.) and biotic stresses adversely affecting the growth and productivity of the crop plants. Further, these stresses as also their destructive potential seem to have substantially increased in the current 'anthropocene' — a geological epoch of 'climate change' caused by human activities. Before the outstanding cytogenetic contributions of Swaminathan to *Solanum* and section *Tuberarium* are elaborated, a brief introduction to the nature of polyploids and the role of polyploidy in speciation would be helpful. These are briefly presented below.

Nature of Polyploids — Observation of a spontaneous doubling of the somatic complement of chromosomes by Hugo de Vries in 1901 in *Oenothera Lamarckiana* was reported as a 'gigas' mutant. The reason for referring to the variant form of the *Oenothera* plant as 'gigas' was that it had much larger flowers and a coarser stem than the normal plants. It was not, however, a case of mutation of alleles (genes), but one of quadruplicating of the entire haploid (n) set of somatic chromosome complement. *Oenothera* plants normally have 14 chromosomes ($2n = 14$) in their somatic cells. However, the variant form had 28 chromosomes, exactly double the number of normal somatic complement of chromosomes. Initial thinking was to regard the 28-form plants as a new species. But genetically, it was just amplification of the genes and chromosomes of the diploid. It had no new genes.

In 1955, Eigsti and Dustin, [18] published their book, *Colchicine in Agriculture, Medicine, Biology and Chemistry* that revealed an alkaloid known as colchicine, extracted from the plant *Colchicum autumnale* inhibits the formation of cell plate (i.e. cytokinesis) after the chromosomes have doubled in number through the cell division called 'mitosis', and consequently the doubled number of chromosomes are 'not' evenly distributed between the two newly formed daughter cells (i.e. one cell becoming two cells with exactly the same number of chromosomes). Instead, the entire complement of doubled number of chromosomes is included in one cell. Successive divisions of the so-formed tetraploid cell results in a tetraploid plant. Due to the formation of quadrivalents (i.e. pairing of four homologous chromosomes in an autotetraploid), the separation and distribution of the chromosomes to the daughter cells are not equal and also not balanced both in numbers and gene content. So, the autotetraploids are largely seed sterile. However, they have substantial vegetative mass and are ideal for fodder and forage crops as well as ornamental plants. Taking advantage of the fact that in fodder and forage crops, the biomass rather than seeds is more important, Swaminathan and his co-workers [19] induced autotetraploidy in *berseem, Trifolium alexandrinum* and *senji* which in English is Indian clover, *Melilotus indica* ($2n = 16$). While the diploids formed eight bivalents regularly in meiosis, the induced autotetraploids formed univalents, bivalents, trivalents and quadrivalents. Their finding that the autotetraploids of sweet clover

(*Melilotus sps.*) did not exhibit any superior agronomic trait in comparison with the parent diploid and were in fact inferior to the diploid in growth and vegetative vigour suggests that there is probably an optimum ploidal level for each species to perform at its best. Their studies throw some light on the fact that autotetraploids are relatively of rare occurrence in nature than allotetraploids, the discussion of which follows in this section.

There are, however, natural autotriploid ($2n = 3x$) plants in several species of food value to humans. These have genomic constitution 'AAA' and have three homologous sets of chromosomes. In the case of beet, an autotriploid has the chromosome constitution of $2n = 3x = 27$. The triploid beet is much larger than the corresponding diploid and contains more sugar, vitamins, etc. However, its meiosis is chaotic. This is because it has three homologous chromosomes of each kind — a_1-a_1-a_1, a_2-a_2-a_2, a_3-a_3-a_3,..., a_g-a_g-a_g. During meiosis, these identical chromosomes tend to make association (pairing) of three chromosomes forming 'trivalents' instead of normal pairing of two homologous chromosomes forming a 'bivalent'. The edible bananas are all triploids (hence they do not form seeds), but they are not autotriploids, but allotriploids resulting from hybridisation among the diploid progenitor ($2n = 22$) species and then subsequent 'triploidisation' through crosses between diploid and autotetraploid forms of the diploid species such as *Musa acuminata*, *M. balbisiana* and *M. paradisiacae*. Autotriploids have also been induced in order to eliminate seed formation in watermelon, first produced in Japan. Triploid apples have high level of preference.

Swaminathan's extensive research with potatoes has involved production of autopolyploids (i.e. autotetraploids $2n = 4x = 48$ and autooctoploids $2n = 8x = 96$) as intermediaries in breeding programmes to improve the quality of the cultivated potato. Their genomic constitutions can be represented as AAAA in the case of an autotetraploid and AAAAAAAA in the case of an autooctoploid.

In nature, autopolyploids occur among fewer species for reasons that these do not combine chromosomes containing complementary gene pools from varyingly evolved with new mutations and natural selection but still related species. Yet, the reason for their spontaneous occurrence in a few species is that the autopolyploids achieve vigour and resilience at the cost of fertility. On the other hand, the polyploids referred to 'allopolyploids' or simply 'alloploids' are substantially more frequent in nature.

The evolutionary pathway of several economically important alloploids (e.g. wheat, cotton, potato, sugarcane, groundnut, etc.) substantiates the fact that several diploid species largely related to each other, but had genetically diverged through spontaneous mutations and natural selection to varying degrees, first hybridise and form the F1 inter-specific hybrids. These inter-specific or even inter-generic hybrids would be mostly sterile due to non-homology of their component chromosomes. Formation of bivalents that involves paring of chromosomes which are homologous to each other would be infrequent, whereas the unpaired chromosomes remaining as univalents would be more commonly observed in the meiotic divisions. Cytogenetic studies in many cases have confirmed the observation of univalents in the first generation of inter-specific and inter-generic hybrids. The evolutionary future of these sterile hybrids depends on the doubling of their chromosome complements (i.e. non-homologous sets of chromosomes). These are illustrated below:

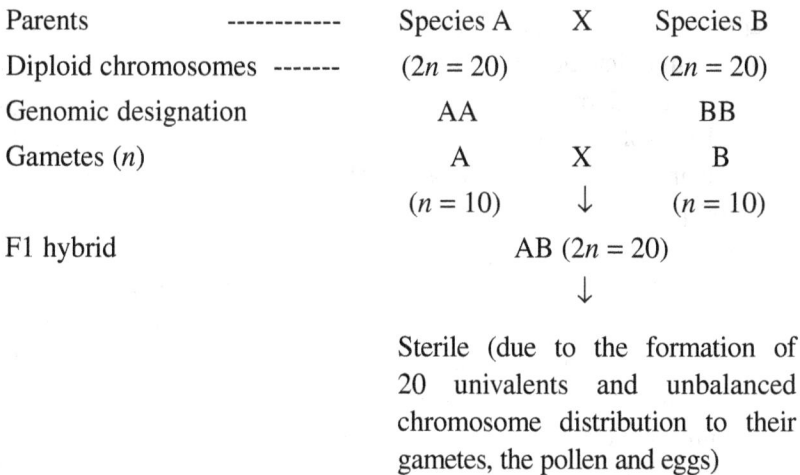

Parents	------------	Species A	X	Species B
Diploid chromosomes -------		$(2n = 20)$		$(2n = 20)$
Genomic designation		AA		BB
Gametes (n)		A	X	B
		$(n = 10)$	$\downarrow$	$(n = 10)$
F1 hybrid			AB $(2n = 20)$	
			$\downarrow$	

Sterile (due to the formation of 20 univalents and unbalanced chromosome distribution to their gametes, the pollen and eggs)

In course of time of the order of thousands of years, there occurs one in millions of meiotic divisions the formation of pollen and egg with unreduced chromosome numbers by chance. The unreduced pollen and eggs have $n = 20$ each. Both the pollen and eggs also contain the full complement of chromosomes of both the A and B genomes. While pollination is at random, it could happen by chance that an unreduced pollen fertilises an unreduced egg. The resultant embryo would then have $2n = 40$ chromosomes comprising two A and two B genomes. Its somatic chromosome complements could be represented as $2n = 4x = 40$ with AABB as the genomic constitution. From the point of pairing of chromosomes in meiosis, 20 bivalents would form. The 10 pairs of homologous chromosomes of A genome form 10 bivalents, and so is the case with the 10 pairs of homologous chromosomes of B genome. This tetraploid is an allotetraploid with regular meiotic pairing of the chromosomes and formation of fertile pollen and eggs.

Wheat (*Triticum sps.*) is of special interest in cytogenetics and it is also a major staple food crop. Wheat revolution in India brought about by Swaminathan and Borlaug in the mid-1960s is widely referred to as Green Revolution. This wheat involved in the Green Revolution is an allohexaploid. Its somatic chromosome complement could be written as $2n = 6x = 42$. This means that the bread (*chappathi*) wheat has six sets of the basic chromosome complement (x) of each with seven chromosomes. The hexaploid wheat cultivated for making *chappathis*, bread, etc., is known as *aestivum* wheat. The hexaploid *vulgare* wheat is also cultivated. There is also an allotetraploid form of wheat, the *durum* wheat which is used for making porridge, South Indian *uppuma*, etc. Its chromosome complement is written as $2n = 4x = 28$. Corresponding to *durum*, there is the *emmer* allotetraploid wheat. The evolution of the *durum* ($2n = 28$) and *aestivum* ($2n = 42$) wheats is fascinating. Since they form 14 and 21 bivalents regularly in meiosis, and they are quite fertile, confirming that these are not autoploids. Extensive research on genome analyses of the polyploid wheats by American, Japanese and British scientists led to the elucidation of the genomic constitution and the probable progenitor species of the *durum* and *aestivum* wheats. The evolutionary pathway and history of the allotetraploid and allohexaploid wheats are as follows.

Approximately, 5,00,000 years ago in the Fertile Crescent, two diploid progenitors *Triticum urartu* (AA genome) and *Aegilops speltoides* (BB genome) hybridised to form the tetraploid species *Triticum dicoccoides* (AABB genome). This tetraploid species was domesticated approximately 10,000 years ago, resulting in *emmer* wheat (*Triticum dicoccum,* AABB) and *durum* wheat (*Triticum turgidum,* AABB). The diploid *einkorn* wheat (*Triticum monococcum,* AA) was domesticated around the same time. *T. monococcum* is a selection several thousands of years ago from its wild form called *Triticum boeoticum. Einkorn* is a German word which literally means 'one kernel' indicating the single (one) kernel per spikelet. *Einkorn* wheat is far more nutritious than the present day tetra-ploid and hexaploid wheats. It has two to four times more of vitamin A, beta-carotene, lutein, riboflavin, etc. Finally, around 10,000 years ago, a tetraploid species (AABB) hybridised with *Triticum tauschii* (DD genome) to form the two hexaploid wheat species *Triticum aestivum* (AABBDD) and *Triticum spelta* (AABBDD). The primary domestication trait in wheat was non-shattering, with secondary traits being glume reduction to improve threshing, changes in plant architecture and in ear and kernel size, loss of seed dormancy, and lower grain protein and increased grain carbo-hydrate content. Today, derivatives of many of these species are used for very specific purposes. For example, the hexaploid *T. aestivum* has been bred for high protein quality needed for bread making, breeding within tetraploid *T. turgidum* has resulted in *durum* wheat used to make pasta, and speciality markets are currently being developed for *einkorn* wheat as an alternative for people suffering from celiac disease. The multiple ploidy levels and distinct market classes are responsible for extensive population structure within wheat.

The genomes of the diploid progenitors of wheat (*T. urartu*) ($2n = 14$) with genomes AA, *A. speltoides*, ($2n = 14$) with genomes BB, and *T. tauschii*, ($2n = 14$) with genomes DD, were neither completely homolo-gous nor completely non-homologous. They all had their component chromosomes varyingly and partially homologous with each other.

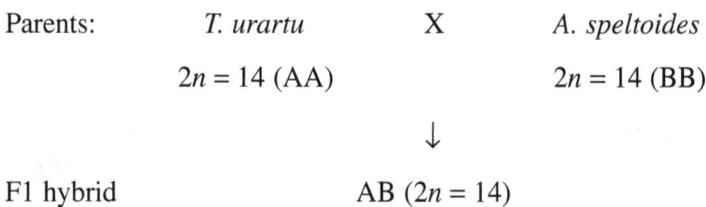

Parents: *T. urartu*   X   *A. speltoides*

$2n = 14$ (AA) $2n = 14$ (BB)

↓

F1 hybrid   AB ($2n = 14$)

Meiosis — If the A and B genomes are completely differentiated in their homologies, 14 univalents are expected in meiosis. However, the AB ($2n = 14$) hybrid forms a few bivalents in addition to univalents. This means partial homology between the chromosomes of the A and B genomes. However, in allotetraploid (AABB), 14 bivalents are observed. The hybrid AB is largely sterile due to 'cryptic structural differences' between the chromosomes of A and B genomes. In such a situation, the tetraploids are expected to form pairing of four 'partially homologous' (or 'homeologous') chromosomes (i.e. Quadrivalents). In the hexaploid wheat, association of six homeologous chromosomes (hexavalents) is expected to form and disturb the meiosis. The consequence would be varying degrees of sterility. However, this has 'not' been the case. The allotetraploid *durum* and the allohexaploid *aestivum* wheats form strictly 14 and 21 bivalents, respectively. The question is as to how these polyploids which have partially homologous (i.e. 'homeologous') chromosomes have become *diploidised*. An interesting discovery by Riley and Chapman [20] in their paper, *Origin of genetic control of diploid-like behaviour of polyploid wheat* was that experimental deletion of a small fragment from the short arm of the 5th chromosome of the B genome results in the formation of multivalents (i.e. quadrivalents, pentavalents and hexavalents) in the hexaploid wheat plant. This discovery suggested that certain gene(s) on chromosome 5B controlled the bivalents formation, notwithstanding the homeology of seven chromosomes each of six sets of genomes. Such marvels of nature are not exactly reproducible in the laboratories. This inference can be illustrated with the human-made alloploid of wheat and rye. The resultant product called *Triticale* combines the genomes of *Triticum durum* ($2n = 28$) or *T. aestivum* ($2n = 42$) and rye, *Secale cereale* ($2n = 14$). Available reports suggest that a hybrid between *T. aestivum* ($2n = 6x = 42$) and *S. cereale* ($2n = 2x = 14$) was first produced over 100 years ago. The hexaploid *Triticale* was produced by crossing tetraploid wheat (*T. dicoccoides*, $2n = 4x = 28$) and diploid rye (*S. cereale*, $2n = 2x = 14$). The F1 hybrid had $2n = 3x = 21$ chromosomes. Denoting the genomes of the tetraploid wheat ($2n = 4x = 28$) as AABB and of diploid rye ($2n = 2x - 14$) as RR, the genomic constitution of the F1 hybrid is ABR, each genome having seven chromosomes. The F1 forms 21 univalents due to lack of homology among the chromosomes of the A, B and R genomes. When the chromosome number of this hybrid is doubled by using colchicine treatment, the allohexaploid ($2n = 6x = 42$) has AABBRR. These allohexaploid *Triticales* have already been produced experimentally and have been in cultivation for over four to five decades. However, unlike the natural allohexaploid wheat (*T. aestivum*), the synthetic allohexaploid *Triticale* is not fully fertile and its meiosis shows the occurrence of univalents indicating some disturbance in meiotic pairing of homologous chromosomes. Over the years, there has been remarkable refinement through selection of the allohexaploid *Triticales* for high fertility and normal meiotic behaviour. Tetraploid *Triticales* were produced more than three decades ago by crossing *T. monococcum* ($2n = 2x = 14$, AA) and rye ($2n = 2x = 14$, RR). The F1 hybrid ($2n = 2x = 14$, AR) was doubled to obtain allotetraploid ($2n = 4x = 28$, AARR) *Triticale*. Yet, even after three decades, meiotic pairing revealed 20–90% univalents [21]. These and several such studies greatly inspire even the highly accomplished scientists.

With the new knowledge gained by Muller's demonstration that X-ray induces mutations in 1927, several plant geneticists initially thought that they had attained the power of the Creator (say the Hindu God 'Brahma'), but soon realised that nature had put a ceiling on the degree of variation that could be induced by radiation and chemical mutagenesis. Swaminathan's analytical papers [22, 23] deal with micromutations, macromutations, systemic mutations, etc., and how the frequency of induced mutations decreases with increasing magnitude of effects. Analysing the free-threshing Q locus in the genus *Triticum*, he concludes that it has evolved through a series of tandem repeats. What this means is that the Q locus that is regarded to have arisen in a single-step macromutation in nature may be obtained through tandem repeats of Q [24]. The purpose of brief reference to these is that 'allotetraploidy' has been an evolutionary mechanism to evolve new species in nature; the scientists, no doubt, have also produced a few allotetraploids experimentally, but these continue to have varying levels of problems in the nature of meiotic abnormalities, sterility, change in taste of food, palatability, etc. These problems are encountered wherever humans try to copy what nature does especially in the biological world.

2.3 Cytogenetic Research in *Solanum*, Section *Tuberarium*: From Wageningen to Wisconsin and Beyond

In 1949, on arrival in Wageningen, Swaminathan boarded with a traditional Wageningen family. His evening meal or the supper was routinely prepared by the landlady and it mostly consisted of vegetables, potato and a little meat which was a luxury in those days in The Netherlands just after the end of the German occupation. As a strict vegetarian then,

Swaminathan was content with potato and vegetables although he slowly accepted scrambled or boiled eggs more to please the landlady than for any other reason. The issue was that landlady felt so sorry for her guest living on potatoes. It is not known whether she knew at all that Swaminathan was also living for improving the potatoes in Wageningen. Living on potatoes and living for potatoes was, of course, an interesting part of his life in Wageningen.

Perusal of the literature on the cytogenetics of potatoes compiled by Swaminathan and Howard [25] reveals that bulk of the voluminous literature published until 1952 dealt mostly with the descriptions of potato varieties from different geographic regions, mutations and selection, occurrence of diseases as well as nematodes-induced damage, photoperiodism, problems in potato breeding, improvement in cytological technique to study the chromosomes in *Solanum*, susceptibility to infection by *P. infestans*, inheritance pattern of several characters, etc. There have been very few cytological and cytogenetic investigations in *Solanum* section *Tuberarium* until early 1950s when Swaminathan made his presence in the field of cytogenetic research of *Solanum* section *Tuberarium*. The Karyotype (i.e. description of the numbers and structure of the chromosomes in the somatic cells) was first published in 1927 by Levitsky and Benetzkaja [26]. In the following decade, 1930s, W. Ellison [25] carried out cytological studies in the British varieties of potato. What can be reasonably surmised is that there had been very few publications on the cytology and cytogenetics of *Solanum* section *Tuberarium* prior to 1950. Further, these publications had wide time gap, which suggests a lack of concerted focus on the cytogenetic research on potato and its wild relatives until the early 1950s, when Swaminathan appeared on the scene.

As stated in Section 2.2, Swaminathan published his first paper on the cytological aspects (i.e. chromosome numbers of cultivated potato and its wild relatives) in *Der zuchter* [17]. This paper laid the foundation for sound cytogenetic approach to elucidate inter-relationships among potato and its wild relatives and the wild diploid and polyploid species. He also pointed out that several diploid ($2n = 24$) wild species have genes for resistance to a variety of pests, diseases, frost, heat, drought, etc., but most of diploid wild relatives ($2n = 14$) do not cross readily with the cultivated tetraploid potato ($2n = 48$) and therefore, additional technique would be needed, that is, the doubling of the chromosome number of the diploid wild species ($2n = 24$) so that the autotetraploid forms ($2n = 4x = 48$) cross more successfully with potato. He had also standardised the method to obtain autotetraploid experimentally. In this paper [17], he also describes the experimental method employed by him to double the chromosome number of the diploid species. He details the production of polyploid plants as follows: *By the use of the drug colchicine, plants with the doubled chromosome number are now easily obtained. By treating seeds or growing tissues of established plants with colchicine in solution or in admixture with agar or lanoline, polyploid sectors or entire polyploid plants can be produced. The colchicine-agar seed treatment method, which has proved to be effective and useful for inducing chromosome doubling in the potato species, is carried out in the following way. Sterilized Petri dishes are filled a third full with equal parts of 0.5% colchicine and 2% agar solution. In a few minutes, the mixture attains a jelly-like consistency and the seeds whose chromosome numbers have to be doubled are spread over it. The Petri dishes are kept at a temperature of $\pm 12°C$ and the seeds germinate within a week. The germinated seeds, when their roots are about 4 mm long, are removed from Petri dishes, washed well in water, and planted in pots. When the first leaves appear, the plantlets are transplanted into soil.*

While the autotetraploids occurring in the colchicine treated group of plants can be initially identified based on general growth habits, leaf thickness and length–breadth index, stomatal size, pollen size, fertility, etc., leaf smears for confirmation of the doubled number of chromosomes in the cells of the basal portion of young leaves were used. It is emphasised that the leaf-smear method using 8-hydroxyquinoline was also developed by Swaminathan in collaboration with Professor Prakken [27]. In the field of scientific research, it is often necessary to standardise the existing techniques or even develop entirely new ones in order to elucidate the structure and function. Swaminathan first perfected the method of inducing chromosome doubling and then ascertaining the somatic chromosome numbers by leaf-smear method in genus *Solanum*, section *Tuberarium*. The simple leaf-smear technique developed for studying the somatic chromosome complement of potato and its wild relatives was later extended to barley shoot-tip cells in the late 1950s and 1960s by Professor Robert Nilan and his co-workers at the Washington State University, Pullman. I and my students have also used it to assess the bridges and fragments of chromosomes in the barley shoot-tip cells exposed to gamma-rays and their post-irradiation modification by radioprotectors such as caffeine, cysteine, etc.

Swaminathan's publication, *Notes on Induced Polyploids in the tuber-bearing Solanum species and their crossability with Solanum tuberosum*, [28] was written after he had joined the University of Cambridge, England, but based entirely on the large amount of research carried out by him during his short stay at the Wageningen University, The Netherlands. This paper brought out the sterility problems in the hybrids between the cultivated potato (*S. tuberosum*, $2n = 48$) and its wild relative species, both diploids ($2n = 24$) and tetraploids ($2n = 48$). An interesting but also intriguing observation reported in the paper is that *S. acaule* ($2n = 48$) does not readily cross with *S. tuberosum* ($2n = 48$), but an autooctoploid of *S. acaule* ($2n = 96$) when used as the female parent more readily crossed and formed F1 *acaule–tuberosum* hybrids. The meiosis of the F1 with uncertain number of chromosomes of *S. acaule* and 24 chromosomes of the potato (*tuberosum*) was expectedly chaotic. However, genes confer-ring resistance to frost could be transferred to the cultivated *S. tuberosum*. Based on a strong foundation of science, Swaminathan's paper in *Nature* [29] demonstrates how he first elucidated the causes of cross-incompatibility among the Mexican diploid species of *Solanum* and then how he devel-oped an *artificial stigma* technique to overcome the cross-incompatibility. Overcoming the problem of cross-incompatibility was a prerequisite to breeding for disease resistance of cultivated potato. Among the serious pathogens destroying the cultivated potato is the disease called 'late blight' caused by pathogenic fungus called *P. infestans*. Interestingly, there are several Mexican species which are resistant to *P. infestans*. The Mexican *Solanum* species comprise diploids ($2n = 24$), triploids ($2n = 36$) and tetraploids ($2n = 48$). There are also a few species having $2n = 60$ (pentaploid) and $2n = 72$ (hexaploid). The crosses between the Mexican species, *S. pinnatisectum* and *S. bulbocastanum* and also those between *S. pinnatisectum* and *S. lanciforme* failed to set seeds. Swaminathan noted these failures were due to 'pollen–stigma incompatibility'. He studied ovaries with style and stigma of *S. pinnatisectum* fixed in a mixture of

Carnoy's solution (alcohol, acetic acid and chloroform, at 6:1:3) after 24, 36, 48, 72 and 120 h after pollination. Then these were sectioned and stained with acid fuchsin and light green. Examination of the slides revealed that in both crosses, there was little or no germination of the pollen on the stigma even after 72 h from the time of pollination. He also found that 72 h is sufficiently long duration for fertilisation to be completed in the pollen–stigma compatible crosses. To overcome the pollen–stigma incompatibility, he made two different approaches: one was to smear the stigmatic exudation from the pollen parent on the stigma of the female parent before pollination; another approach was to apply a suitable artificial medium for pollen germination on the cut surface of the style after removal of the stigma. An intermediate method adopted by Swaminathan led to the formation of berries and seeds in both the crosses, viz., *S. pinnatisectum* X *S. bulbocastanum* and *S. pinnatisectum* X *S. lanciforme*. Methodology was that the stigma, with a small portion of the style, was removed with a pair of fine scissors from the flowers of *S. pinnatisectum,* and a drop of an agar–sucrose–gelatin medium was applied on the decapitated surface of the style. The medium was prepared by dissolving 0.5 gm of agar, and 2.5 gm of sucrose in 25 cc of distilled water to which 0.5 gm of gelatine was added. The pollen grains of many species of *Solanum* were found to grow well in this medium. After applying pollens to this medium on the cut surface of the style, the cut style was covered with a piece of moist cotton wool. From the crosses *S. pinnatisectum* X *S. lanciforme* and *S. pinnatisectum* X *S. bulbocastanum* thus made, 4 and 3 berries containing an average 39 and 8 seeds, respectively, were obtained. Seeds grown from these crosses using artificial stigma were confirmed as true inter-specific hybrids by Dr. R.W. Hougas of the University of Wisconsin, USA. Seeds from these crosses were grown at the Inter-regional Potato Introduction Station, Sturgeon Bay, Wisconsin in the summer of 1954. This technique which came to be called as the 'Artificial Stigma Technique' paved the way to transfer the gene responsible for resistance to late blight from *S. bulbocastanum* to the cultivated *S. tuberosum.* Interestingly, *S. bulbocastanum* is not only highly resistant to late blight but also probably the only species then known to be immune to late blight. While this piece of work had been carried out in Wisconsin in 1954, his doctoral research at the University of Cambridge, England

(1950–1953), was extremely thorough and comprehensive in elucidating the genome affinities (i.e. homology of the chromosomes) of potato and its wild relatives. He also produced autoploids of some of the species by doubling their somatic chromosome numbers by using colchicine treatment and amphidiploids of the F1 inter-specific diploid hybrids.

Starting as a beginner of cytogenetic studies in *Solanum* section *Tuberarium* at the Wageningen Agricultural University in 1950, Swaminathan emerged as a leader in the field within the next 4 years. The intellectual calibre and diligence of most young researchers during the 1950s were certainly above average; in this background, it is absolutely commendable that Swaminathan was far ahead of his times then and continues to be so even now in his 91st year of age. In this regard, his paper with his Professor Prakken [30] at Wageningen is among the foremost, if not the first, to demonstrate the formation of unreduced eggs formed in nature in the inter-specific hybrids. When different species within a genus or even different forms of the same species are crossed with each other, plants with an increased number of chromosomes are sometimes formed, e.g. triploids from crosses between two diploids, tetraploids from crosses between diploid and tetraploid and so on. In these cases, the unreduced germ cells generally function. The unreduced germ cells (eggs, pollen) are known to arise either by obliterating meiosis or by meiotic disturbances in the first or second division by which the uninucleate embryo sacs or pollen grains get the unreduced number.

In collaboration with Professor R. Prakken, Swaminathan analysed five different inter-specific F1 hybrids as follows:

1. *S. demissum* ($2n = 72$) X *S. caldasii* ($2n = 24$)
   - The F1 hybrids had $2n = 48$ as expected; no increase in the chromosome number.

2. *S. demissum* ($2n = 72$) X *S. chacoense* ($2n = 24$)
   - Found six F1 hybrids with $2n = 48$ as expected
     Two of the hybrids had $2n = 60$, a pentaploid; unreduced male gamete (pollen) with $n = 24$ fertilised the reduced egg ($n = 36$).

3. *S. tuberosum* ($2n = 48$) X *S. phureja* ($2n = 24$)
   - Three hybrids with $2n = 36$ as expected.

4. *S. phureja* ($2n = 24$) X *S. tuberosum* ($2n = 48$)

    -   Three hybrids with $2n = 36$ and three hybrids with $2n = 48$ chromosomes were obtained. The female parent contributed unreduced egg ($2n = 24$).

5. *S. tuberosum* ($2n = 48$) X *S. chacoense* ($2n = 24$)

    -   Of five hybrids studied, two had $2n = 36$ (as expected) and three had $2n = 48$ chromosomes. Unreduced male gamete (pollen) had fertilised egg with $n = 24$ chromosomes.

The above-mentioned studies confirm that in the spontaneous evolution of polyploids, especially the allopolyploids, formation of unreduced eggs and pollen is a natural phenomenon. It had been previously speculated by Stebbins 1947 [31], but it had not been demonstrated until 1952 by Swaminathan.

Another significant contribution by Swaminathan to the understanding of species inter-relationships among the cultivated and wild diploid and polyploidy species was published in 1952 [32]. *S. demissum* is a hexaploid species ($2n = 72$). It is highly resistant to blight. A cross between *S. demissum* and the cultivated *S. tuberosum* ($2n = 48$) would result in an allopentaploid ($2n = 60$). Instead, the idea of Swaminathan was to cross *S. demissum* ($2n = 72$) with *S. rybinii* ($2n = 24$) to get a tetraploid hybrid with $2n = 48$ and then to cross this hybrid with *S. tuberosum*. In fact, data on the formation of bivalents, quadrivalents and univalents (which show the extent of regular meiotic pairing and fertility) obtained in actual experiments favoured Swaminathan's approach. Data (configuration at first metaphase in *S. demissum* hybrids) taken from Table 3 of Ref. [32] are reproduced in Table 2.1.

The average number of berries and average seeds per berry in the cross numbers 1–3 are shown in Table 2.2.

The implications of these observations for both basic and applied goals are quite substantial. On the basic aspect, occurrence of increased frequencies of bivalents than quadrivalents and univalents suggests varying degrees of homologies among the constituent genomes. Divergence of homologies due to progressive cryptic structural differences has, in fact, been known in the cytogenetics of polyploids.

Table 2.1.   Configuration at First Metaphase in *S. demissum* Hybrids.

Hybrid and Cross Number	Chromosome Number ($2n$)	Meiotic Quadrivalents (4)*	Pairing Bivalents (2)*	Univalents (1)*
1. *S. demissum* X *S. tuberosum*	60	0.187	24.5	10.252
2. *S. demissum* X *S. rybinii*	48	0.09	22.9	1.84
3. *S. demissum* F1 X *S. tuberosum*	48	0	23.33	1.33

*The numbers within the parentheses show the number of chromosomes in meiotic pairing.
Source: Table 3 of Ref. [32].

Table 2.2.   Number of Berries and Average Seeds per Berry in *S. demissum* Hybrids.

Hybrid and Cross Number	Number of Berries	Number of Seeds Per Berry
Cross No. 1 *S. demissum* F1 X *S. tuberosum*	4	61
Cross No. 2 (*S. demissum* X *S. rybinii*) F1 X *S. tuberosum*	6	91

In the cases of extreme divergence of the genetic content and structural homologies of the chromosomes between the cultivated potato and a wild species with desirable genes, use of intermediary bridging species was yet another brilliant idea demonstrated by Swaminathan. This formed part of his doctoral research at the University of Cambridge, England.

In 1953, Swaminathan with his Professor H.W. Howard published a masterly review on *The Cytology and Genetics of the Potato (Solanum tuberosum) and Related Species* [25]. This continues to be a valuable source material even after six decades. The cytological studies starting from 1899 until 1926–1927 reported the chromosome number of the cultivated potato varyingly as $2n = 36$, $2n = 34$, $2n =$ about 48, etc. It was only through the study of meiosis, the gametic number (n) was reported correctly as 24, corresponding to the number of chromosomes as $2n = 48$ in the somatic cells. One of the most useful compilation in the review paper [25] is the list of most of the *Solanum* species, their chromosome numbers in the somatic cells, taxonomic group and the locality of type specimen. The list includes 118 species (diploids, triploids, tetraploids, pentaploids, hexaploids) from 13 different series. These species are from different places such as Bolivia, Peru, Mexico, Argentina, Uruguay, Ecuador, Chile, Venezuela, Costa Rica, etc.

So far as homology of the chromosomes of the diploid species ($2n = 24$) is concerned, these are close enough to form 12 bivalents in meiosis. Data for eight intra-series hybrids and 15 inter-series hybrids compiled by Swaminathan are as follows:

I. Intra-series hybrids (all species have $2n = 24$)

1. *S. rybinii*   X   *S. stenotomum*
2. *S. goniocalyx*   X   *S. rybinii*
3. *S. phureja*   X   *S. rybinii*
4. *S. geniocalyx*   X   *S. stenotomum*
5. *S. geniocalyx*   X   *S. bukasovii*
6. *S. simplicifolium*   X   *S. rybinii*
7. *S. macolae*   X   *S. simplicifolium*
8. *S. caldasii*   X   *S. chacoense*

II. Inter-series hybrids

9. *S. rybinii*   X   *S. emmeae* and reciprocal
10. *S. catarthrum*   X   *S. emmeae* and reciprocal
11. *S. kosselbrenneri*   X   *S. emmeae*
12. *S. phureja*   X   *S. jamesii*
13. *S. verrucosum*   X   *S. chacoense*
14. *S. verrucosum*   X   *Papa chusa*
15. *S. polyadenium*   X   *S. jamesii*
16. *S. chacoense*   X   *Papa chusa*
17. *S. henryi*   X   *S. verrucosum*
18. *S. henryi*   X   *S. vavilovii*
19. *S. parodii*   X   *S. infundivuliforme*
20. *S. rybinii*   X   *S. verrucosum*
21. *S. rybinii*   X   *S. chacoense*
22. *S. rybinii*   X   *S. caldasii*
23. *S. rybinii*   X   *S. commersonii*

However, the formation of 12 bivalents in the meiosis of all the above-mentioned F1 hybrids did not lead to fully fertile plants with normal seed set. The F1 hybrids of *S. henryi* X *S. chacoense* and *S. henryi* X *Papa chusa* were so weak that they died before flowering. The F1 hybrids *S. verrucosum* X *S. chacoense* and *S. verrucosum* X *Papa chusa* were self-sterile, but gave good yields of seeds when pollinated with other hybrids. The lessons learnt are of fundamental interest to genetics and possibly what we know today as epigenetic mechanisms which operate beyond the pairing of homologous chromosomes and formation of bivalents in their meiotic divisions.

The general effects of doubling of chromosome number of different *Solanum* species with $2n = 24$, $2n = 36$ and $2n = 48$ also varied considerably. The octoploid *S. tuberosum* ($2n = 96$) had larger stomata, larger flowers with calyx more hairy and were somewhat dwarfed, and these were late in flowering. The seed set was highly reduced. On the other hand, the doubling of the chromosome number of the tetraploid species ($2n = 48$), *S. acaule* and *S. longipedicellatum* did not result in discernible changes.

Production of polyploids of various diploid and tetraploid species in *Solanum* is not just of academic interest. The applied interest is to enhance the crossability of experimentally produced polyploids with the cultivated potato (*S. tuberosum*) and transfer the desirable genes for resistance to a variety of biotic and abiotic stresses to it from its wild relatives which possess these genes. Several wild species such as *S. rybinii* ($2n = 24$), *S. kesselbrenneri* ($2n = 24$) and *S. polyadenium* ($2n = 24$) do not easily cross with *S. tuberosum* ($2n = 48$). However, the same diploids after their chromosome numbers are doubled ($2n = 4x = 48$), cross appreciably with *S. tuberosum*.

Swaminathan's analyses of the nature of polyploidy species in *Solanum* in general, and that of *S. tuberosum* in particular are indeed remarkable. Emphasising on the sound observation that in potato, any set of 12 chromosomes can pair with any other set, he rules out the possibility of defining the cultivated potato either as an autotetraploid or a strict genomic allotetraploid. Partial homologies of the genomes of the different diploid species suggest that the cultivated potato could be what is called 'segmental allopolyploid'. Despite large parts of the different sets of chromosomes being homologous, cryptic genic/structural differences preferentially promote pairing of the totally homologous chromosomes, resulting in mostly bivalents, in the tetraploids. In support of this contention, Swaminathan [33] refers to two of the amphidiploids, *S. yabari* X *S. stenotomum* and *S. parodii* X *S. saltense* and notes that these have a high frequency of quadrivalent formation. He has reasoned that there are no small (cryptic) structural differences

Table 2.3.   Frequency of Quadrivalents/Cell.

Material	Autotetraploid	Allotetraploid
Case 1		
S. chacoense (female parent)	3.67	—
S. rybinii (male parent)	3.45	—
F1. S. chacoense X S. rybinii	—	None
Case 2		
S. verrucosum (female parent)	4.56	—
S. macolae (male parent)	3.93	
F1. S. verrucosum X S. macolae (male parent)	—	1.08

between their chromosomes. These interpretations are so original, brilliant and indeed led to subsequent experimental confirmation consequent upon the advances in knowledge and techniques to unravel the '*hidden* or *not easily resolvable factors*' in chromosome homology among related species.

Some of the unpublished data from Swaminathan's research at the University of Cambridge are tabulated in the review paper published in *Bibliographia Genetica* [25]. These are of interest from the point of assessing the homology of the component genomes in an autotetraploid and the allotetraploids. Some of these are given in Table 2.3.

Table 2.3 is only illustrative and not exhaustive. Swaminathan had studied several *Solanum* species for their formation of quadrivalents (pairing of four homologous chromosomes) in their autotetraploid (AAAA) state and in their amphidiploids (AABB) form. Formation of 'zero' quadrivalents in the F1 amphidiploid means that homology between the genomes of the two parent species is almost non-existent, whereas the occurrence of a few quadrivalent in the F1 amphidiploid suggests partial homology between the genomes of the two parents. Based on his extensive studies, Swaminathan provided experimental evidence to the view of Stebbins [31] that the two categories of polyploids, viz. autopolyploids and allopolyploids usually recognised are not sharply distinct from each other, irrespective of the criteria on the basis of which distinction between them is drawn.

Genetic studies in the *Solanum* species have been far more challenging. Among a few reasons, one is that most of the potato varieties are heterozygous (Aa, Bb are the alleles) than being homozygous (AA, aa,

BB, bb). Further, the genetic segregation in F2, F3, etc., are vitiated by varying percentages of degenerate seedlings. Yet another problem is that the hybrids often do not flower, and exhibit varying degrees of sterility.

Inheritance of resistance to Late blight (*P. infestans*) was once thought to follow the Mendelian monohybrid ratio of 3:1 (susceptible:resistant). Soon this was found to be erroneous and efforts put in by several geneticists led to a postulate of a series of R genes. In this series, R1 was assigned a resistance value of 1, R2 of 2, R3 of 3, R4 of 4. The genotypes with regard to susceptibility or resistance to the late blight in a given variety or species could comprise R genes with different resistant values such as 1,2,3 and 4 (Table 2.4).

In all probability, it is for the first time that a cytogeneticist, Swaminathan has used the disomic and tetrasomic inheritance in a *Solanum* hybrid to conclude that the quadrivalents and bivalents in the tetraploid are not formed in a 'random manner' but result from pairing among 'specific chromosomes'. In his paper, *Disomic and Tetrasomic Inheritance in a Solanum Hybrid* published in *Nature,* Swaminathan [34] analysed the mode of inheritance of three loci in a cross between *S. macolae* and *S. simplicifolium* both having $2n = 24$ chromosomes. The *simplicifolium* parent used in the cross had simple leaves, a prominently winged stem and white corolla, in contrast to the compound leaves, rudimentary stem wing and purple

Table 2.4.   Resistance (R) Value and Incidence of Late Blight in Potato.

Genotype*	Sum of Values	Reaction — Resistant or Susceptible	Segregation Ratio in Selfed Progeny R:S
$R_1R_1R_3R_4$	9	S	1:35
$R_1R_2R_3R_4$	10	S	3:33
$R_2R_2R_3R_4$	11	S	5:31
$R_2R_2R_4R_4$	12	R	12:24
$R_3R_3R_3R_4$	13	R	23:13
$R_3R_3R_4R_4$	14	R	27:9
$R_4R_4R_4R_4$	15	R	36:0

* Plants with a R value of 12 and above are resistant to late blight. Much of the inheritance studies pertaining to various traits (e.g. flower colour, shape of tubers, etc.) are compiled in a manner of providing ready reference to potato breeders.

corolla possessed by the *macolae* parent. Except for the corolla colour in the *macolae*, the two species bred true for their characteristics. The *macolae* parent was heterozygous for 'P', a dominant factor necessary for the formation of purple colour in the corolla. By studying the F1 and pattern of segregation of the three traits, Swaminathan determined the genotypes of *S. macolae and S. simplicifolium* parents as LL WW PP and ll ww pp (L, for compound leaf; W for prominent stem wing and P for purple corolla), respectively. In F2 progenies from F1 plants with compound leaves, prominent stem wing and purple corolla, there was monogenic segregation for all the characters. Then, he produced tetraploid plants from the seeds from initial crosses by treating them with colchicine. Thus, he had tetraploid hybrids. These tetraploid hybrids produced a mean frequency of 1.19 quadrivalents, 0.095 trivalents, 0.909 univalents and 21.025 bivalents. The number of quadrivalents per cell ranged from 0 to 4. The tetraploid plants were fertile. These tetraploid hybrids were selfed so as to obtain the segregating generations C2, C3 and C4 (corresponding to the F2, F3 and F4 plants of the diploid hybrid). He observed that all the plants in the different generations had broad stem wings like C1 and *simplicifolium* parents. Since the C1 plants would have had the genotype WWww for stem wing character, the lack of segregation in the subsequent generations would mean that the *simplicifolium* chromosomes carrying factor W were pairing preferentially at the tetraploid level. The segregation ratios for leaf type, however, were typical of tetrasomic inheritance; simplex and duplex genotypes were identified by the 3:1 and 35:1 ratios obtained on selfing C3 plants. In order to make sure that the simple leaf character in them is not due to the loss of chromosomes carrying the 'L' factor, he checked the chromosome numbers and found that all the plants had $2n = 48$. Thus, he demonstrated that the 'L' in tetraploid *macolae* X *simplicifolium* segregated tetrasomically. So far as the corolla colour was concerned, the C2 progenies (on selfing C1 or sib-crossing C1) segregated in a ratio of 3 coloured to 1 white. The genotypes of the C1 plants with purple and white colours were determined as PPpp and pppp, respectively. From the ratios observed, Swaminathan concluded that the chromosomes in which the 'P' locus is situated pair preferentially at tetraploid level.

Those were the days when there were no computer-linked sophisticated instruments to print out the data that could directly go into an original research paper. Extremely patient and assiduous bench work in the laboratories and then a really great analytical mind to draw valid conclusions from huge amount of data were the hallmark of excellence. Then a publication of the work in *Nature* authenticates the generation of new knowledge. The extensive basic work carried out by Swaminathan led to the transfer of genes for frost resistance from *S. acaule* to the cultivated potato and development of the 'Alaska Frostless Potato'. Swaminathan, it is noted, had been focussing his studies on *S. polyadenium* ($2n = 24$) because of its blight-resistance with field immunity to viruses A and X.

Swaminathan continued his vigorous research in the cytogenetics of potato in the Department of Genetics, College of Agriculture, University of Wisconsin, USA. He published the work done there with his Professor Hougas [35]. *S. verrucosum* is a diploid with $2n = 24$; it forms 12 bivalents in meiosis. Its variant, *S. verrucosum var. spectabilis*, is a hexaploid ($2n = 72$), and it regularly forms 36 bivalents in meiosis. Cytogenetic relationship between (i) diploid varieties of *S. verrucosum* and *spectabilis* and (ii) *spectabilis* and *S. demissum* was studied. The diploid *S. verrucosum* strictly formed 12 bivalents per cell as expected. The *S. verrucosum var. spectabilis* ($2n = 72$) formed 35.8 bivalents and 0.3 univalent per cell. The F1 hybrid of *S. spectabilis* X *S. demissum* ($2n = 72$) formed 21.5 bivalents, 17.7 univalents, 3.2 trivalents and 0.4 quadrivalents per cell. The reciprocal of this hybrid, i.e. *S. demissum* ($2n = 72$) X *S. spectabilis* ($2n = 72$) formed 22.6 bivalents, 25.2 univalents, 0.5 trivalents and 0.1 quadrivalents per cell. Discussing these observations from taxonomic point of view, Swaminathan and Hougas [35] pointed out that variety *spectabilis* of *S. verrucosum* does not behave as a variety of *S. verrucosum* ($2n = 24$) in terms of chromosome number, plant habit, corolla colour and shape and berry shape. The view then prevalent was that all chromosome races are best recognised as distinct species even if morphological differences are slight. A novel approach by Swaminathan was that *spectabilis* could be regarded as a variety of *S. demissum* ($2n = 72$) since *spectabilis* also has the same chromosome number ($2n = 72$) and both *verrucosum* and *spectabilis* belong to the same Mexican taxonomic series. *S. demissum*, a Mexican species used widely in potato breeding because of its resistance to Late blight disease is the only hexaploid species described under section *Tuberarium*. He also emphasises the observation that F1 hybrids of *spectabilis* and *demissum* have a mean number of 17–25 univalents per cell at metaphase I of meiosis and they are pollen and ovule sterile. Based on these observations, the conclusion was that at least one set of 12 chromosomes of *spectabilis* and *demissum* differ from each other. The result is contrary to the generally held view that there is no chromosome differentiation among the species of *Tuberarium*. The conclusion of the authors, under these considerations, is that *var. spectabilis* is a distinct species and it is named *S. spectabilis* (correll) by Swaminathan and Hougas. It may be recalled that Swaminathan started his cytogenetic research in potato and its relative wild diploid and tetraploid species as a predoctoral researcher in 1949–1950 at the Wageningen University, The Netherlands. In just 1 year, he collected valuable data which resulted in seven publications in reputed international journals such as *Zuchter, American Potato Journal, Euphytica, Genetica,* etc. Five of them were authored by Swaminathan. Two of them were published jointly with his Professor Prakken. During his days as a doctoral research scholar at the Plant Breeding Institute, School of Agriculture, University of Cambridge, England (1950–1952) and Post-doctoral Research Associate (1952–1953) in Genetics in the Department of Genetics, University of Wisconsin, USA, Swaminathan had carried out a lot of original research providing enough data to publish about a dozen papers in leading international journals such as *Nature* (two papers authored by Swaminathan), *Euphytica, Genetica, American Potato Journal, Bibliographica Genetica, Journal of Heredity, Genetics* and *Genetica Agraria.* The number of pages of all these publications are about 500 pages with excellent illustrations of chromosome numbers, meiotic pairing behaviour of chromosomes in diploids, polyploids, the various hybrids and their amphidiploids, morphology of the plant types, pollen sterility, etc. The depth of cytogenetic investigations and the wide spectrum of inclusion of a large number of diploid and polyploid species from different taxonomic series covered in his studies is absolutely outstanding and his contribution remains unmatched to this date.

Two of his papers containing data of his research at the University of Wisconsin, USA, have major conclusions and these are in fact guideposts to future researchers and potato breeders. Towards the end of his 5 years of thorough investigation of the cytogenetics of potato and its wild diploid and polyploidy relatives, Swaminathan published two papers, one in *Genetics* [33] and another in *Journal of Heredity* [36] in which he elaborates microsporogenesis in 12 commercial varieties of potato and draws conclusions on the nature of polyploidy in 48 chromosome species of the genus *Solanum*, section *Tuberarium*, respectively. The chromosomes of these species are of small size, and there was no good technique available to obtain good preparations of microscopic slides containing cells with well-stained and well-spread out chromosomes until Swaminathan developed the so-called 'hydroxyquinoline smear method' and published it [27]. Prior to it, there seems to have been a general aversion to take up the challenge of cytological and cytogenetic studies in *Solanum*, section *Tuberarium*.

Swaminathan's paper [36] on the microsporogenesis of 12 different commercial potato varieties has brought out the following:

(i) The commercial potato varieties with $2n = 4x = 48$ chromosomes do not always form 24 bivalents in meiotic pairing. The bivalents formed are in the range of 13.84–20.56 per cell in the different varieties. The frequency of quadrivalents varied from 1.12 to 4.40 per cell. Trivalents were noted in the range of 0.40–1.0 per cell. Univalents occurred in the range of 0.98–2.14 per cell. Swaminathan [36] pointed out that the trivalents and univalents could be the result of 'desynapsis' (i.e. premature disjoining from an association of two or more pairs of chromosomes) of chromosomes from quadrivalents. The significant aspect of this paper is that it suggests a progressive shift from formation of quadrivalents in a tetraploid (whether it is an auto or allo) towards formation of more bivalents. This means that quadrivalents in fewer numbers naturally occur. Only in truly strict allotetraploids, in which the constituent genomes AA and BB are already sufficiently differentiated, bivalents are more regularly formed. These are the genomic allopolyploids. Between the strictly autotetraploids (AAAA) and strictly genomic allotetraploids (AABB), there are two other categories of polyploids — namely the segmental allotetraploids (AAA′A′) and autoallopolyploids (AAAABBBB). The segmental allotetraploids result from chromosome doubling of the F1 hybrid whose female and male parents are not either completely homologous as in the case of autotetraploids or fully heterologous as in the case of genomic allotetraploids; actually, the chromosomes of the genomes of the two parents of segmental allotetraploids are partially homologous as shown below:

Parents	Species A	X	Species A'
	$(2n = 24)$	$\downarrow$	$(2n = 24)$
F1		$\rightarrow$	$(2n = 24)$ (AA')
F1 doubled (amphidiploid)		$\rightarrow$	$2n = 48$ (AAA'A')

A and A' are partially homologous, so that they can form a few bivalents in the F1 hybrid. Yet, the F1 plant would be sterile to varying degrees because of small (cryptic) structural differences between the chromosomes of the two genomes A and A'. With the doubling of their chromosome numbers, the pairing will be between those of AA and A'A'. So, it would form 24 bivalents, and the fertility is substantially restored. The genomic constitution of a segmental allotetraploid will be AAA'A' $2n = 4x = 48$.

Stebbins [31] has shown that autoallopolyploids occur only at the octoploidy $(2n = 8x)$ level and higher. This is because an autoallopolyploid should have a genomic constitution of AAAABBBB. Assuming that A and B genomes comprise 12 chromosomes each, the autoallopolyploid is an octoploid with $2n = 8x = 96$. Its meiosis is expected to have a maximum of 24 quadrivalents; however, octavalents and other associations could possibly occur depending on the inter-homologies of the chromosomes of A and B genomes.

With this background, the noteworthy contribution of Swaminathan in two of his above-mentioned papers could be summarised:

The cultivated potato $(2n = 48)$ originated as an autotetraploid. Therefore, its genomic constitution could be written as AAAA. This conclusion, however, raises a fundamental question regarding the lack of formation of 12 quadrivalents. In this regard, Swaminathan's papers [33, 36] clarify that one of the 12 commercial varieties in his study showed nine quadrivalents in a cell and the number of quadrivalents per cell was 4.80. In comparison, a colchicine-induced autotetraploid of *S. polyadenium* ($2n = 4x = 48$) formed 5.56 quadrivalents per cell [37]. So, it is clear that the induced autotetraploid, even its first (C1) generation, does not form 12 quadrivalents. Swaminathan explains that an autopolyploid may, soon after its origin, gradually shift to a diploid behaviour both genetically and cytologically. Further, he points out that the chromosomes of the autopolyploid plants may undergo differentiation through structural changes like inversions, and interchanges, gene mutation and hybridisation which will lead to a change from a multivalent to bivalent type of synapsis [36]. In support of his point of view, there are several papers on progressive diploidisation of chromosome pairing in autopolyploids as well as segmental allopolyploids. In the earlier section, reference was made to the genetic locus in the '5B' chromosome of the 'segmental' allotetraploid *durum* wheat ($2n = 4x = 28$) and 'segmental' allohexaploid *aestivum* wheat ($2n = 6x = 42$) [20]. The discoverer was Ralph Riley and the year of reporting of the discovery of '5B' — locus that diploidises the meiotic pairing in polyploid wheats was 1961. It is hence obvious that Swaminathan had discussed genetic and cytological mechanisms of diploidisation of meiotic behaviour in autopolyploids, and segmental allopolyploids as early as 1954. Swaminathan's statement that is so comprehensive is reproduced below [36].

A shift to a bivalent type of synapsis will be favoured in seed propagated autopolyploid plants both by natural and human selection if there is a reduction in seed fertility as a result of multivalent formation. Human selection will be especially operative if the plant part of economic value is the seed. Such a shift, may, however, be of relatively little importance for purposes of selection in vegetatively propagated plants like potato. He goes on, *There is also evidence of a negative correlation between pollen and seed fertility and tuber yield in S. tuberosum. If this is so, pollen and ovule sterile plants as well as poor and non-flowering ones would have a selective advantage at least under human selection. Many of the commercial varieties are in fact pollen sterile. Hence, the chromosome associations in current varieties of S. tuberosum may be more representative of the situation that existed in this species at the time of its origin than similar associations observed in varieties of a seed-propagated polyploidy species.*

Based on extensive analyses of the meiotic behaviour, he also suggested that (a) *S. acaule* and *S. longipedicellatum* are segmental allopolyploids, (b) *S. andigenum* and *S. tuberosum* had autopolyploidal origin though the commercial varieties can be considered as segmental allopolyploids and (c) both on taxonomic and cytogenetical grounds *S. andigenum* can be treated as a sub-species of *S. tuberosum*.

By all measures of a highly rigorous evaluation system, Swaminathan's five-year tryst with potato and its wild relatives was vastly productive as never before and not so far after. He became a legend in science to elucidate polyploidal origin of cultivated potato and its wild diploid and polyploid relatives and predicting their future destiny as well. All these do not mean that he was far too deeply engrossed in the polyploids of *Solanum* only. While he was doing his doctoral thesis work at the University of Cambridge, he was invited to make a contribution on 'polyploidy' to *New Biology* edited by M.L. Johnson and Michael Abercrombie, Penguin Books, London. His paper [38] on *Polyploidy and Plant Breeding* discusses the nature of polyploids in general, and elaborates the effects of polyploidy on specific crops such as barley, sugar beet, fodder crops, vegetables and ornamental crops. The major revelation from this paper [38] is that he has been as much focused on the applications and benefit to humankind from the research on polyploids as he was on satisfying intellectual curiosity regarding their origin and evolutionary significance.

2.4 Swaminathan and Polyploidy — Beyond Wisconsin Back in India

His biographers Iyer [4] and Dil [1] have referred to Swaminathan's decision to return to India in early 1954 in spite of a very handsome offer of a teaching-cum-research professorship given to him in the Department of Genetics, University of Wisconsin. Most people in his place would have just grabbed it; instead, he decided to return to India to develop appropriate science-based technological initiatives to achieve self-sufficiency in food production. On his return to India, there was no vacancy of a scientist's position for him. In those days, there were very few jobs in agricultural research and development. Notwithstanding his highly impressive research accomplishments and scientific leadership already evinced by him, he could get only a temporary position as Assistant

Botanist at the Central Rice Research Institute (CRRI), Cuttack, Orissa. This was a brief assignment, yet its outcome was stupendous since it was there that he sowed the seeds of the First Green Revolution of India. At the CRRI, Cuttack, he was assigned to research and development of ferti-liser-responsive hybrids of *indica–japonica* varieties of rice *(Oryza sativa)*. It indeed paved the way for the much needed acceleration to yield gains.

In October 1954, he joined the Botany Division of the Indian Agricultural Research Institute (IARI), New Delhi, as Assistant Cytogeneticist, after having been selected by the Union Public Service Commission, Government of India. He obviously continued his research interests in potato amidst the several other research works and teaching responsibilities assigned to him. Particularly, two of his outstanding papers published in *Nature* (London) 1955 and 1956, respectively [29, 34], exemplify his leadership in science. Further, he initiated interesting programmes of research to elucidate whether polyploidy influences the genesis of the leguminous root nodule and its mechanism. His research on how polyploidy influences the radiosensitivity of plants is yet another revelation of his intellect and innovative breakthroughs in science. He also kept his ideas and research goals focused on elucidating the Origin of Indian and European varieties of cultivated potatoes and the probable centres thereof [39]. There have been several new dimensions of polyploids which have been conceived and successfully investigated by him and his students. These are briefly discussed in the following sections:

(a) Polyploidy and the genesis of the leguminous root nodule

Since 1930s, there are reports associating polyploidy with nodular stimulation/nodular infection of the leguminous root nodules by rhizobia, the nitrogen-fixing bacteria belonging to Gram-negative group. In fact, the occurrence of polyploid chromosome numbers in nodule cells of many leguminous plants was almost established and it was clear that a 2:1 ratio is invariably maintained between the chromosome number of infected and normal cells irrespective of whether the plant is initially a diploid or a polyploid. Hence, the question was to resolve the nature of association between polyploidy and the differentiation of the leguminous nodule. At that time, the proposal of Wipf and Cooper [40] was that disomatic cells are possibly essential for the release of the rhizobia from the threads as well as for tissue proliferation culminating in nodule formation. On the other hand, in several plants, infection by bacteria and viruses had been found to induce polyploidy in the affected cells. Thus, there was a difficulty in differentiating between 'cause' and 'effect' while judging the exact relationship between nodule genesis and the sporadic occurrence of disomatic cells in root tips of the leguminous plants. It is at this juncture that Swaminathan, with one of his brilliant students Dr. Bhaskaran, designed appropriate experiments to establish whether polyploidy was the 'cause' or the 'effect' of rhizobial infection in the nodules of the leguminous plants. This epoch-making paper by Bhaskaran and Swaminathan [41] was published in 1958. Their experimental design was absolutely ingenious and far ahead of the tools and techniques available over six decades ago in the 1950s. Their experimental approach had the following components:

(i) Diploid and colchicine-induced autotetraploid *berseem* (*T. alexandrinum*) and *senji* (*M. indica*) were used as the experimental materials. Both of them have somatic chromosome number as $2n = 2x = 16$ and their autotetraploids have somatic complement of chromosome number as $2n = 4x = 32$. Earlier, Swaminathan along with two of his fellow scientists [42] had shown that autotetraploids of *berseem* and *senji* differed prominently only in cell size and their conclusion was that *senji* has probably attained a cell size optimum for the species even at the diploid level.

(ii) The strategy of Bhaskaran and Swaminathan [41] was to assess the DNA content of the nuclei of root tip cells, root tissue from differentiated region and nodules. It involved cytophotometric measurement of the Feulgen dye-stained nuclei at two carefully selected wavelengths. The photoelectric microspectrometer that was set up by them is reproduced here from Ref. [41] (Fig. 2.1). The DNA measurements of the resting and dividing nuclei in the root meristem

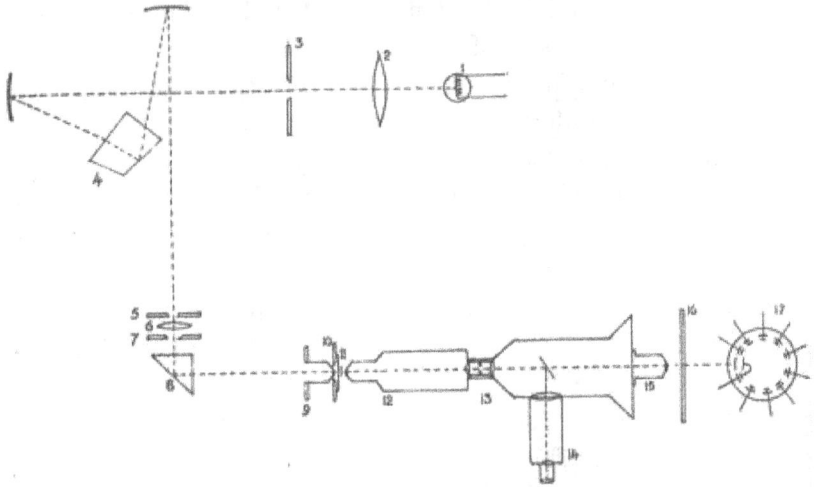

Figure 2.1.   Photoelectric microspectrometer.

Source: Reproduced from Ref. [41], pp. 77–86.

showed a relationship of 1:2. The nuclei from tetraploid *berseem* gave double the value found in the diploid. Some nuclei in the mer- istematic tissue showed a value intermediate between $2x$ and $4x$ classes; the authors rightly suggested that these were probably in early or mid-interphase. In this context, it is mentioned that in the late 1950s and early 1960, the interphase was resolved into 'G1' (pre-DNA synthetic gap), S (the DNA synthetic phase), 'G2' (the post-synthetic gap) phases before metaphase sets in. Thus, it could be noted how their accurate estimation of the DNA content in the 'resting' (G0-stage) and in the dynamic 'G1–S–G2–M' inter- phase greatly supplements the conclusion of Howard and Pelc [43] on the DNA synthesis in cells. In a subsequent paper, Bhaskaran and Swaminathan [44] have more elegantly analysed the stage of DNA synthesis during mitosis and further extended it to meiosis as well. Their paper put to rest varied views then prevalent that DNA synthe- sis occurred in anaphase, late telophase, etc. Their data unequivo- cally demonstrated that DNA content doubled from 2C to 4C content during interphase in mitosis of rye, *S. cereale* ($2n = 14$). In meiosis, their data in *S. cereale* showed that DNA content increased from 2C to 3C just prior to leptotene and reached 4C content by pachytene. These data of Swaminathan and his student have stood the trial of time until today — about 60 years later.

As in the diploid, the nuclei from the root meristem of tetraploid *berseem* fell into three categories, viz., $4x$ (resting nuclei), $8x$ (dividing nuclei), and an intermediate class (interphase nuclei). In the *infected cells* of the root nodules, the DNA values corresponded to three classes — $2x$, $4x$ and $8x$ in diploid *berseem* and $4x$, $8x$ and $16x$ in tetraploid *berseem*. Thus, the failure of the chromosomes to the two daughter cells at anaphase is the mechanism of chromosome numbers doubling in the nodule cells. Data provided by the authors in Table 2 of their paper [41] are reproduced below to show that chromosome doubling in the nodule cells is a certain feature (Table 2.5).

Also, the total number of nodules per plant was calculated in diploids and autotetraploids of *berseem* and *senji*. In berseem, the tetraploid had a significantly higher number of nodules than the diploids, while in *senji*, the tetraploid had a slightly but not significantly more number of nodules in comparison with the diploid. Relevant part of the data extracted from Table 3 of their paper [41] is provided below (Table 2.6):

The cytochemical and the cytological data [41] established that (i) a 2:1 relationship generally exists between the chromosome number of the nodule and root tip cells, (ii) besides cells with double the somatic chromosome number, normal diploid cells as well as cells with a higher

Table 2.5. Chromosome Numbers in Nodule Cells.

Material	Number of Cells with $2n =$		
	16	32	64
berseem			
Diploid	13	121	1
Tetraploid	—	7	150
senji			
Diploid	3	78	—
Tetraploid	—	12	82

Source: Reproduced from the data provided by the authors in Table 2 of Ref. [41].

Table 2.6.   Number of Nodules in Diploid and Tetraploid Plants.

Material	Mean Number of Nodules on			Total Number of Nodules
	Tap Root	Primary Root	Secondary Root	
berseem				
Diploid	86.5	44.4	11.2	142.1
Tetraploid	114.2	145.4	35.2	294.8
senji				
Diploid	37	27.6	11.2	75.8
Tetraploid	40.9	31.2	12.5	84.6

Source: Data extracted from Table 3 of Ref. [41].

polyploidy chromosome number also occur in the nodules, (iii) mitotic aberrations such as spindle inhibition and arrested cytokinesis are noted in the nodule cells, and (iv) polyploid cells occur but rarely in the differentiated root tissues, there being no correlation between nuclear size and DNA content in them. These findings were in agreement with those of a couple of earlier authors who found that the chromosome number in the infected nodular cells in diploid and naturally occurring or induced tetraploid strains of leguminous species is in each case twice that characteristic of uninfected somatic cells of the same plant. However, there was no evidence for the occurrence of disomatic cells of spontaneous origin in the root tissue either prior to or after the initiation of nodule formation. Swaminathan's leadership in this field of science has unequivocally demonstrated that the *bacterial infection occurs first, and this leads to spindle inhibition in metaphase–anaphase border of the mitosis, and consequently, the chromosome doubling is the effect and not the cause of the bacterial infection.* This conclusion remains valid even today.

(b) Studies on effects of ionising radiation in crop plants

Before expounding Swaminathan's magnificent research to elucidate the difference between polyploids and their corresponding diploids in response to ionising radiation, a brief introduction to the chromosomal and genetic effects in the irradiated cells and organisms is presented as it would be extremely useful. Discovery of ionising radiation towards the end of the 19th century and the early years of the 20th century has resulted in epoch-making impacts on biology, medicine and agriculture. Briefly stated, X-rays were discovered in 1895 by Wilhelm Roentgen, gamma-rays in 1896 by Henri Becquerel, alpha particles in 1899 by Ernst Rutherford, beta particles in 1900 by Henri Bequerel and neutrons in 1932 by James Chadwick. From a biological point of view, currently, the basic unit of radiation dose (energy) absorbed per unit of tissue is called Gray (Gy) that represents deposition of 1 J of energy per kilogram of tissue. It is physical unit for expressing the amount of the ionising radiant energy absorbed in biological systems. About 50–60 years ago, this term had not been introduced, and therefore, the terms used by Swaminathan in his papers were initially the 'R' for Roentgen Units ('R' defined as 2.58 X 10^{-4} C/kg or 1 C/kg = 3876 R) and later 'rad' (radiation absorbed dose which is that 1 rad = 0.01 J/kg of tissue). The letter 'C' stands for calorie. One 'Gy' is equivalent to about 100 rads.

Exposure to ionising radiation and radioactive material has been integral to life itself. Very sophisticated studies have now established beyond any iota of doubt that daily exposure to certain minimum level of ionising radiation is absolutely essential for healthy and productive life. If that minimal level of exposure to ionising radiation is denied, morbidity is the result. This aspect had remained obscure because almost all the experimental studies conducted during the years between 1900 and 1950s involved moderate to high doses. It is even true that low and high doses of ionising radiation induce beneficial and deleterious effects in biological systems [45].

Until about 1920, the major biological effects of X-rays and gamma-rays, often noted, consisted of skin reddening (erythema), cataract of the eyes at moderate doses (say up to 600 R delivered in chronic exposure doses). However, after the atomic bomb detonations over Hiroshima and Nagasaki in August 1945, there was renewed interest in the paper, *Artificial transmutation of the gene* in which Muller [46] had unequivocally demonstrated that X-rays at moderate–high doses induce lethal mutations (later found to be deletion of small segments of the X-chromosome) in the fruitfly *Drosophila melanogaster.* The major cause of unwanted anxiety and overestimation of likely health hazards (mostly cancers) was Muller's postulate that ionising radiation effects are stochastic, that even very very low doses could cause deleterious genetic changes. He was awarded the Nobel Prize for Medicine in 1946 (the year after atomic bombing of Hiroshima and Nagasaki) and he became a dominant figure in the *United Nations Scientific Committee on the Effects of Atomic Radiations* (*UNSCEAR*) in Vienna, Austria. At his instance, the UNSCEAR adopted the *linear no-threshold* (*LNT*) model to describe the relationship between dose and detrimental genetic consequences. It is emphasised that LNT model has not received experimental evidence so far. In fact, the recent studies on molecular and cellular events induced by ionising radiations challenge the LNT model.

About the same time as Muller was studying the genetic effects of X-rays in fruitfly, L.J. Stadler, also from USA, was doing the same, but with maize and barley. Stadlar's results [47] also showed that X-rays and radium rays induce genetic effects in maize and barley. The demonstration that ionising radiations induce mutations in crop plants, led Ake Gustafsson of Sweden to write a long review of about 100 pages on the scope of utilising induced mutations for improving agricultural plants. This paper [48] has an extensive bibliography of 147 titles. The extensive bibliography, however, did not make any distinction in the radiobiological effects between diploid and polyploid species. It is in this context that Swaminathan's pioneering studies on the altered radiosensitivity or radioresponse of polyploid plants has become highly relevant since many crop plants requiring improvement through plant breeding are polyploids. Since polyploids have more than two sets of genomes and also more than a pair of alleles (alternate forms of expression of gene(s)) and moreover, their gametes which participate in the fertilisation are 'polyhaploids', it is likely that the effect of a *mutated recessive allele* (more than 80% of the induced viable/visible mutations are recessive in expression) could be masked by the presence of a 'normal' (non-mutated) allele in the corresponding homologous or homeologous chromosome. While this is a relevant and important question of applied interest, there are equally pertinent considerations basic to radiobiology and cancer radiotherapy. In this regard, the reference to 'oxygen effect' in the irradiated diploids and polyploids by Swaminathan in a few of his papers is of interest. These are discussed in this section.

Swaminathan's pioneering studies on how polyploidy influences the sensitivity of plants to ionising radiation are the first to shed light on some of the most intriguing questions. Further, they opened avenues for right type of experimental designs for achieving greater insight into radiobiological mechanisms in the diploid and polyploid systems.

The year was 1956, and Swaminathan had just then been elevated as the Cytogeneticist (from Assistant Cytogeneticist) in the Botany Division of the IARI, New Delhi. In India, unless one occupied higher position with greater rights (and of course responsibilities) especially in the scientific institutions of the government, it is quite difficult to initiate newer lines of research programmes, irrespective of the urgent need and immense possibilities for crop improvement associated with them. In this context, I remember an episode. When Swaminathan was the Cytogeneticist, and not yet the Head of the Division of Botany of the IARI, New Delhi, he set up laboratories equipped to use fruitfly (*D. melanogaster*) and human peripheral leucocyte (blood) cultures *in vitro* for basic research in genetics and cytogenetics. He was questioned by his senior officers as to the purpose of wasting resources and time over organisms which are of no agricultural value. With his brilliance coupled with pleasing manner, he convinced his seniors that much of the then knowledge and basic principles in genetics have come from research using *D. melanogaster* and the Botany Division had also teaching obligations to the Post-graduate School of the IARI. About the human leucocyte culture laboratory, he pointed that only a few months earlier, a paper on spreading and staining the human chromosomes had appeared, which would eliminate much of the uncertainties regarding the exact number of chromosomes in normal human females and males and that India should also develop expertise in the latest technique. He pointed that the new technique unambiguously determined the chromosomes number in the humans as $2n = 46$; the human males have 44 autosomes + 1X and 1Y chromosomes, whereas the females have 44 autosomes + 2X chromosomes. He also emphasised that a 2-credit course in Human Genetics was being offered by the Division to the students majoring in Genetics and Cytogenetics. Thus, he satisfied the bosses who had little appreciation for wider and deeper understanding of related issues.

The aforementioned deviation from the focal theme of polyploidy and radiosensitivity is relevant to the context of Swaminathan's setting up radiation sources including a 'Gamma *garden*' with 200 Curie cobalt-60 source and initiating basic research with a view to understand the radioresponse of different crop plants under varying physical, chemical and biological conditions so as to achieve a reasonable control over the spectrum and frequency of induced mutations. The Gamma garden set up by Swaminathan in August 1960 nearly coincided with his 35th birthday. The 200 Curie source centrally located in a circular garden was designed to chronically deliver radiation doses, decreasing in dose with increasing distance between the source and placement of the seeds/vegetative propagules in pots or soil (Fig. 3.1). His idea was to overcome *diplontic selection* which is the progressive elimination of few mutated cells by the more rapidly dividing normal (non-mutated) cells in the actively dividing (mitotic divisions) meristematic tissues.

Swaminathan's paper [49] jointly with Dr. S.M. Sikka the then Head of the Division of Botany, IARI on *Fifty Years of Botanical Research at the Indian Agricultural Research Institute* reveals that prior to 1955, there had been no mutation breeding and related basic research in radiation biology. Hence, without doubt, the major programme of Mutation breeding in the Division of Botany of the IARI was first initiated by Swaminathan. I and many other close associates of Swaminathan know full well that inertia between a 'thought' and 'action' is absolutely nonexistent in his case. Within a year of his joining the Division of Botany, IARI as an Assistant Cytogeneticist in 1955, he published a couple of significant papers on the radiosensitivity of the chromosomes of *Einkorn* ($2n = 14$), *Emmer* ($2n = 28$) and *Aestivum* ($2n = 42$) wheats irradiated with fast neutrons (from the Department of Atomic Energy facilities in Mumbai) and particularly the localised chromosome breakage in the *einkorn* (*T. monococcum*) ($2n = 14$). In their paper entitled *Fast Neutron Radiation, and Localised Chromosome Breakage*, Swaminathan and Natarajan [50] showed that the region near supernumerary constriction of satellite II chromosome of *T. monococcum* was preferentially disposed to breakage by fast neutron radiation. This demonstration of a non-random 'clastogenic' (i.e. chromosome breaking action) action of ionising radiation was contrary to the expected *random* action. They also extended the

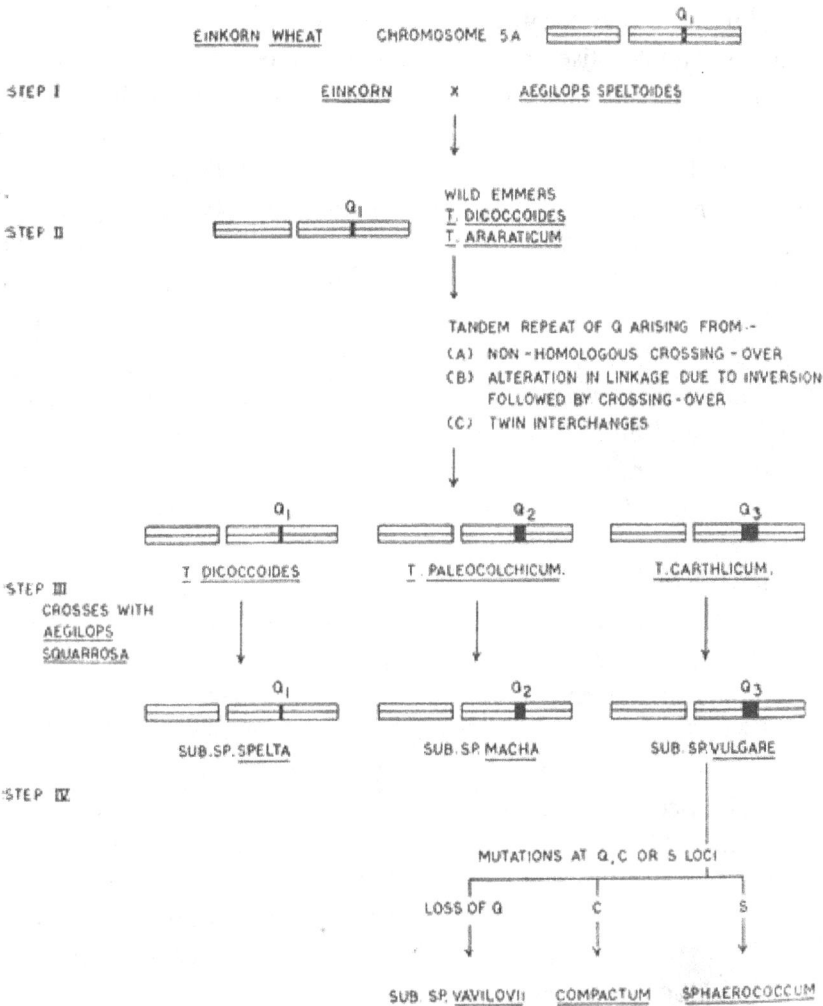

Figure 2.2.   Possible mode of origin of Q of various strengths [76].

studies by including *Emmer* and Bread wheats [51]. These findings were
not only intriguing but also raising questions on the physical/chemical
nature of the supernumerary constriction of satellite chromosome II of *T.
monococcum.* Unlike X-rays and gamma-rays, which induce ionisation
events sparsely distributed along the length of the chromosomes at random, the fast neutrons are densely ionising, and their energy is deposited

in one site or closely localised clusters. Even more intriguing was their finding that the localised breakage appeared to occur in only one of the pair and not at random in either of the two homologous satellite II chromosomes. The conclusion of the authors is quoted as it best explains the intriguing observation: *In case the non-random breakage is confined to only one of the Satellite II chromosome pair, as seems likely from our data, it may suggest the presence of sub-microscopic differentiation between the homologous pair of chromosomes at the vulnerable region. In a similar study carried out in the same variety of T. monococcum using beta radiation from* ^{32}P*, we found no evidence for a preferential susceptibility to breakage of any segment in either satellite II or the other chromosomes. Thus, it would appear that the localize breakage caused by fast neutron radiation is probably a correlated sequence of the specific properties of the neutron particles and the concerned chromosome segment.*

Since 1956, radiation biology involving sparsely and densely ionising radiations has advanced quite significantly. Inter-cellular communication from the ionising radiation-hit cells to the adjacent unaffected cells is now known to result in phenomena called 'bystander effects', 'genomic instability', etc., leading to DNA repair, or if the damage is rather too extensive to repair, to 'cell death' by necrosis, 'mitotic catastrophe', apoptosis, etc. I have discussed these in a review paper [45], and can state that 'linear energy transfer' (LET) matters a great deal. Swaminathan and Natarajan [50, 51] have elegantly alluded to this fact by their phrase *specific properties of the neutron particles and the concerned chromosome segment.*

An outstanding contribution to basic radiobiology by Swaminthan and Natarajan [52] is their paper entitled, *Effect of ultraviolet (UV) pre-treatment on yield of mutations by X-rays in wheat.* The essential findings briefly are as follows:

The most prevalent view then in the 1950s was that radiation effects are cumulative in biological systems and therefore a dose of irradiation delivered at a point of time and another at varying intervals afterwards would result in damage caused by the first and second doses. Accordingly, a review paper on this subject by Muller [53] referred to a study in *Aspergillus* in which combined UV and X-rays induced more mutations than the sum for the two radiation treatments applied separately. In their

Table 2.7. Frequency of Occurrence of Chromosome Aberrations in X-ray and UV + X-ray Treatment (*T. aestivum*).

X-ray Dose (R)	Chromosome Aberrations Mean ± (Standard Error)	
	Direct X-ray Treatment	UV-Pretreatment* + X-rays
11,000	1.11 ± 0.21	0.70 ± 0.13
16,000	1.68 ± 0.31	0.927 ± 0.25
22,000	2.90 ± 0.39	1.53 ± 0.24
33,000	3.6 ± 0.46	1.76 ± 0.38

* UV pretreatment for 1 h did not induce any chromosome aberrations.
Source: Reproduced the relevant part of the data from Ref. [52].

studies, Swaminathan and Natarajan [52] used the hexaploid wheat ($2n = 6x = 42$) as the test system. The UV pretreatment was for 1 h. The UV-pretreatment alone did not induce any chromosome damage, nor any chlorophyll and visible mutations. Following the pretreatment, the seeds were exposed to X-rays at 11,000 R, 16,000 R, 22,000 R and 33,000 R. In the current system of units, these approximately correspond to 110 Gy, 160 Gy, 220 Gy and 330 Gy.

Reproduced in Table 2.7 are relevant part of their data which are self-explanatory.

The authors have also presented data on the mutation frequencies among the progeny of the X_1 plants. The X_1 plants are those raised from the seeds exposed to UV, X-rays and UV + X-rays. The data on mutation frequencies in the progeny of X_1 are similar to the data on chromosome aberrations for doses of exposures 11,000 R and 16,000 R; however, at higher radiation doses (22,000 R and 33,000 R), the viability of the X_1 plants is greatly reduced and the mutation frequency is vitiated. UV-pretreatment increases the survival of the X1 plants and the mutation rate per X_1 plant progeny (%) increases.

From the point of basic radiation biology, the authors had unequivocally demonstrated what in the 1980s was recognised as the phenomenon of *radioadaptive response*, that a relatively smaller dose of a radiation, called the *priming dose* given to cells *in vitro* and *in vivo* organisms significantly reduces the biological damage induced by a subsequent higher and challenging dose. The credit for the discovery of radioadaptive response should really have gone to Swaminathan and his first doctoral student Natarajan, but it went to Sheldon Wolff [54]. I with one of my students (Zeba Farooqui) [55] demonstrated the same but in *in vivo* mice; unfortunately, they had not known the work of Swaminathan and Natarajan [52]. However, by strange coincidence, Dr. Zeba Farooqui got a post-doctoral assignment under Professor A.T. Natarajan in the early 1990s at the Strahlenzentrum of the University of Leiden, The Netherlands. She worked with Professor Natarajan for about 2 years. All these seem to be incredible coincidences.

Their pioneering observations contradicted the assumption that radiation doses delivered at different times are cumulative. Discovery of *radioadaptive response* [54] has already exerted notable implications for elucidation of the mechanisms of radiation damage and is being currently debated in the context of maximum permissible doses for regulatory purposes.

(c) Polyploidy and radiosensitivity

A major need was to study the effect of ionising radiation in the polyploid species with more than '$2n$' chromosomes and more than '2C' DNA to resolve the question whether these are more resistant or susceptible than their corresponding diploid species. So far as the scientific literature in the field of polyploidy and radiosensitivity is concerned, I could only find that it was Swaminathan who had ever made concerted efforts to assess the influence of *polyploidy on sensitivity to mutagens* — not just physical but also the variously acting chemical mutagens. There have been very few studies before Swaminathan and his students, particularly A.T. Natarajan and S. Bhaskaran had started concerted studies on polyploidy and radiosensitivity. It will not be incorrect to infer that there were in fact no studies at all linking the factors governing the radiosensitivity of chromosomes and genes of diploids and polyploids with the recovery of frequency and spectrum of mutations for crop improvement until the appearance of the first paper of Swaminathan [56] on this aspect. This paper was quickly followed by a series of papers in several scientific journals of international repute. One of the most remarkable new knowledge provided by their research as early as 1956 was that the frequency of interaction of the ionising radiation-induced chromosome breaks with each other to form *exchange* type of aberrations depends both on ploidy and the ionisation density (i.e. LET) of the radiation [50, 57]. These studies of basic interest have been essential to understand mechanisms with a view to achieve reasonable control over them and use them in enhancing the production of 'clean mutations' (i.e. without accompanying cryptic chromosome aberrations, sterility, etc.) in crop plants and preferential killing of the hypoxic malignant cells in cancer radiotherapy. With particular reference to mutation breeding in crop plants, Swaminathan [58] has enhanced the awareness of the enlightened lay citizens and the policy makers of the great potential that atoms offer towards food security. He has dealt not only with induction of desirable mutations not readily available in the wild relatives of crop species but also in using ionising radiation for the extension of readily perishable produce such as fresh fruits and vegetables, fish, etc., as well as disinfestation of grains for prolonged storage. He has also often emphasised that an understanding of the advantages and disadvantages caused by the polyploid state with reference to the induction of mutations is also of great practical interest since many important crop plants like bread wheat, potato, upland cotton, tobacco, banana, groundnut and sugarcane are polyploids. The goal was to overcome the disadvantages of polyploidy in mutation breeding and not to feel helpless about it. Stadler [47] was sceptical about the utility of undertaking mutation research in polyploid plants for the reason that it would be difficult to mutate several copies of a gene in the same direction. The question was whether there is any character of value that is controlled just by one gene alone, i.e. one pair of alleles in polyploids. Ake Gustafsson [48] in his exhaustive review has presented data in bread wheat which reveal that enhanced ability of polyploids to tolerate chromosome structural changes may render it more, and not less, favourable for being used in mutation experiments and mutation breeding.

The outstanding treatise in two parts on *Polyploidy and Radiosensitivity in Wheat and Barley* published by Swaminathan and one of his outstanding students Bhaskaran [59, 60] have brought together all the major issues and developments pertaining to the cytology and cytochemistry of diploid and polyploid wheat and barley, exposed to low LET X-rays and high LET neutrons. Most of the papers of the leading radiation biologists of those days were not only cited but also appropriately commented upon. Making appropriate comments on the major problems and perspectives of the research papers published in an incipient area of science (radiation cytogenetics and radiation genetics were new branches in those days) required enormous data derived from well-designed experiments and a masterly command over the physical, chemical and biological events triggered by the absorption of radiant energy from low and high LET radiations in cells and organisms having diploid and tetraploid chromosome numbers as well as 2C and 4C DNA content. These authors also deal with the complexity of the presence of oxygen during and immediately after exposure of the biological systems to X-rays. Their papers refer to the papers of the then top authorities (J.C. Mottram, J.M. Thoday and J. Read) on the radiobiological 'oxygen effect', and bring in their own data and logic to discuss the 'dual role' of oxygen. While oxygen enhances the radiobiological damage induced by X-rays to a greater extent in the diploids than in the polyploids, it is also a fact that oxygen is needed for post-irradiation recovery. It is again emphasised by me that various mechanisms of repair of the DNA damage induced by UV and ionising radiation were only beginning to be understood. Unfortunately, a few untenable assumptions led to faulty experimental designs and wrong conclusions. I have shown how invalid the extrapolation of the model of repair of the UV-induced DNA damage is when low LET gamma-rays and not the UV has been the radiation source [61].

Today, about 60 years after the publication of the paper *Polyploidy and Radiosensitivity* [57], there is still reverential recognition of the sound foundation laid by Swaminathan and Natarajan to our correct understanding of inter-relationships among polyploidy, ionising radiation of low and high LET values. In a later paper, Swaminathan brings in 'oxygen effect' in relation to the radiosensitivity of the polyploids and diploids. In their paper [57], the choice of *T. monococcum* ($2n = 14$), *T. dococcum* ($2n = 28$) and *T. aestivum* ($2n = 42$) was ideal to assess the radiosensitivity in relation to ploidal levels. The radiation sources were X-rays (low LET) and fast neutrons (high LET). Swaminathan and Natarajan determined the total lengths of the chromosome complement of diploid, tetraploid and hexaploid wheats in comparable preparations and these were 172.2, 282.0 and 353.2 μ, respectively. This approach facilitated the estimation of the number of chromosome aberrations induced by X-rays and fast neutrons per 100 μ. Their data showed that

X-rays and fast neutrons induced aberrations (exchanges + deletions) in the diploid, tetraploid and hexaploid wheats as shown in Table 2.8.

Table 2.8. LET, Ploidy and Frequency of Chromosome Aberrations in *Triticum* Species [56].

Radiation	Ploidy	Aberrations Per Cell (Exchanges + Deletions) Mean ± S.E
X-rays	Diploid	0.35 ± 0.087
X-rays	Tetraploid	0.57 ± 0.137
X-rays	Hexaploid	1.11 ± 0.212
Fast Neutrons	Diploid	0.498 ± 0.079
Fast Neutrons	Tetraploid	1.460 ± 0.169
Fast Neutrons	Hexaploid	9.700 ± 0.517

These data corrected an earlier notion that chromosomal radiosensitivity per unit length is identical in haploids, diploids and higher ploids. The quality of the radiation (i.e. the LET) is a major factor as also the level of ploidy. When these data are transformed into chromosome aberrations per 100 μ length, there is a striking revelation in Table 2.9.

Table 2.9. LET, Ploidy and Chromosome Aberrations per 100 μ Length.

Ploidy Level	Chromosome Aberrations Per 100 μ Length	
	X-rays	Fast Neutrons
Diploid	0.20	0.29
Tetraploid	0.20	0.51
Hexaploid	0.31	2.74

The authors point out that the advantage of ploidy in the hexaploids irradiated with fast neutrons is offset by nearly 9–10-fold increase in the frequency of chromosomal aberrations. Swaminathan [56] discussed the phenomenon of the relative biological effectiveness (RBE) of different ionising radiation to explain the higher-than-expected frequency of chromosome aberrations per unit length. Fast neutrons deposit much of the energy at short distances in close proximity. When energy is deposited in close clusters, the proportion of irreparable damage steeply increases. The DNA lesions close to each other interact to form complex aberrations beyond the means of known DNA repair systems. These cells are removed by 'programmed cell death' (apoptosis).

Contrary to the then prevalent views that polyploids would not be suitable material for mutation breeding using ionising radiation, Swaminathan concluded thus: [56] *The outlook for the induction of useful mutations in polyploid plants appears hopeful.* He goes on to say that *The nature of polyploidy with reference to individual loci, the type of function associated with the loci concerned, the genotype of the material, the type of mutagen used, the stage and method of treatment and various pre- and post-irradiation conditions would all influence the frequency and spectrum of mutations obtained. Polyploid plants have in such experiments the advantage of survival notwithstanding the intra-genic and intra- and inter-chromosomal changes that may be induced by the treatment.*

The Second United Nations International Conference on the Peaceful Uses of Atomic Energy was held in July 1958 at Geneva. Swaminathan [62] made a presentation on the *Polyploidy, Radiosensitivity and Mutation Frequency in Wheats* and was greatly appreciated. It was an excellent review of the work done by various eminent, internationally acclaimed mutation breeders and also of research carried out by him and his team of workers at the IARI, New Delhi. In this presentation, he described chlorophyll mutations which are lethal and viable mutations some of which are of useful and practical interest. Towards the end of his presentation at the International Conference, he concluded that the outlook for solving several specific breeding problems in bread wheat through induction of mutations appeared extremely hopeful. Although it was a science-based analysis, it seemed prophetic. He had also included data from his earlier studies which showed that UV-pretreatment significantly reduced both chromosomal aberrations as well as the frequency of chlorophyll and visible mutations in bread wheat. As discussed earlier, he could have been in the current UNSCEAR discussions had he used the term *radioadaptive response*. However, those who might read this book would come to know that the phenomenon of *radioadaptive response* had been observed long before

1980s when Sheldon Wolff and his co-workers coined the term *radioadaptive response* and developed a new school of thought [54].

Among many useful mutations obtained by treatment of the two bread wheat varieties (N.P. 799 and N.P. 809) with X-rays as well as phosphorus-32 (beta rays) and sulphur-35 (beta rays) occurrence of 'awn' (the awns of 1–2 cm in length and confined to uppermost spikelets) mutations are not rare. Interestingly, the paper published in *Nature* by Swaminathan and his co-workers [63] showed that awn mutations significantly enhanced the yields of the mutant forms of both N.P. 799 and N.P. 809. They cite a paper [64] in the literature to suggest that in warm and dry areas, awns play a particularly important part in grain production.

(d) Genetic component in radiosensitivity

The genetic basis for radiosensitivity has been investigated as early as 1960s by Swaminathan and his co-workers including the doctoral research scholars working under his supervision. The Lethal Dose for 50% survival in about 10–15 days (LD-50) for the disomic ($2n = 6x = 42$) bread wheat (variety Chinese Spring) was 31,620 R. In their paper, the authors Jagathesan and Swaminathan [65] produced monosomics (deletion of one chromosome from the genome $2n = 6x = 42$, and consequently, the plants have $2n = 41$ chromosomes). They produced monosomic plant for each of the 42 chromosomes belonging to the three homologous sets each of seven pairs of chromosomes. With this array of monosomics, they were able to assess any abrupt decrease in the dose required to cause lethality to 50% of the irradiated population within 7–10 days. In other words, the loss of a particular chromosome sharply decreasing the LD-50 meant that it had genetic factors for increased radioresistance. Their finding was that the D genome chromosomes, when dispensed with, enhanced the radiation sensitivity to a much lower degree that those belonging to the A and B genomes. The monosome for chromosome 6B showed the highest radiation sensitivity.

In an earlier paper, Swaminathan and Pai [66] had discussed the 'differential radiosensitivity among the probable genome donors of bread wheat'. It had been mentioned by Swaminathan and Natarajan [57] that the chromosome breaks induced in bread wheat irradiated with fast neutrons was far in excess of the number to be expected on the basis of comparison with *T. monococcum* ($2n = 14$) and *T. dicoccum* ($2n = 28$). Further,

Bhaskaran and Swaminathan [59, 60] found that while diploid and tetra-ploid wheats show a radiation response similar to those of diploid and autotetraploid barley, the hexaploid bread wheat is consistently more sensitive particularly to densely ionising (high LET) neutrons. In order to find out if one of the diploid progenitors i.e. *T. monococcum* ($2n = 14$, AA genomes), *Aegilops squarrosa* ($2n = 14$, DD genomes) and *A. speltoides* ($2n = 14$, BB genomes) is possibly more radiosensitive, they carried out appropriate studies. Both chromosomal aberrations and percentage of X_2 families segregating for chlorophyll mutation were used as parameters.

Briefly reproduced are the data of Pai and Swaminathan (Tables 2.10 and 2.11) [66].

Table 2.10.   Frequency of Chromosome Aberrations in X_1 Generation.*

Species	X-ray Dose (R units)	Mean Number of Breaks Per Cell ± SE
T. monococcum	11,000	0.775 ± 0.07
T. monococcum	16,000	0.905 ± 0.07
A. speltoides	11,000	0.69 ± 0.19
A. speltoides	16,000	0.735 ± 0.07
A. squarrosa	11,000	0.81 ± 0.12
A. squarrosa	16,000	1.05 ± 0.07

*X_1 is the irradiated generation.
Source: Reproduced from Ref. [66].

Table 2.11.   Frequency of Chlorophyll Mutations.

Species	X-ray Dose (R units)	% Age of X_2 Families Segregating for Chlorophyll Mutations
T. monococcum	Control	0/(2,200)
T. monococcum	11,000	22.2 (316)
T. monococcum	16,000	16.6 (263)
A. squarrosa	Control	0 (10)
A. squarrosa	11,000	51.8 (211)
A. squarrosa	16,000	36.4 (130)

Values within the parenthesis are the number of X_1 plants scored.
Source: Reproduced from Ref. [66].

In the above studies, mutation data for *A. speltoides* could not be obtained due to complete failure of seed setting in this species in New Delhi. The authors bring out an interesting fact. Though all the three species used in the present study have $2n = 14$, the total chromosome length is much less in *A. squarrosa* in comparison with the other two species. In an earlier paper, Swaminathan with Natarajan [57] and with Bhaskaran [60] have shown that the ratio of total chromosome length in diploid, tetraploid and hexaploid wheats is not of the order 1:2:3 but is about 1:1.6:2. The data of Bhaskaran and Swaminathan [59] concerning the DNA content in *T. monococcum, T. dicoccum* and *T. aestivum* varied in the ratio 1:1.45:1.92. Their earlier research paper [67] to this effect was published in the prestigious journal *Experimental Cell Research*. The total length of the chromosome complements of *T. monococcum, A. speltoides var. ligustica* and *A. squarrosa* varied in the ratio of 1:1.239:0.710 [66]. The important point is that the total chromosome length and DNA content of varieties of bread wheat ($2n = 6x = 42$) are only twice (and not thrice as to be expected) as great as that of *T. monococcum* ($2n = 2x = 14$). The inference is that chromosome diminution has occurred in the tetraploid and hexaploid wheats during their evolution. With reference to enhanced radiosensitivity of the bread wheat, Swaminathan and his students conclude that *the squarrosa genome might have introduced into the hexaploid bread wheat a physiological state which promotes the incidence of a higher frequency of aberrations and visible mutations following treatment with mutagens.*

Scientists while performing experiments towards gaining insight into specific issues or towards solving specific problems often come across observations that are unexpected or are unintended. Many of them do not pursue them as they believe that these fall outside their major scientific pursuits and would therefore digress from their main research focus. Swaminathan, on the other hand, would engage himself and also motivate his students to probe the new and unexpected result to its logical end. Thus, the fact that there has been diminution of chromosome length in tetraploid and hexaploid wheat led Swaminathan to design appropriate experiments to get to know more about it. His efforts resulted in the paper, *Chromosome diminution and evolution of polyploid series in Triticum* [68]. These scientists undertook a critical study of the length of the chromosome complement, karyotype symmetry and DNA content of nuclei in the three progenitor diploid species of the bread wheat as well as the tetraploid and the hexaploid species. As of 1961, the three diploid progenitor species of the polyploidy wheats were as follows:

(i)   *T. monococcum* or *T. aegilopoides* ($2n = 14$) — AA genomes
(ii) *A. speltoides* ($2n = 14$) — BB genomes
(iii) *A. squarrosa* ($2n = 14$) — DD genomes

Their data on chromatin length and DNA content in *Triticum* and *Aegilops* species and synthetic amphidiploids are reproduced in Table 2.12.

These data confirm that *T. monococcum*, *T. dicoccum* and *T. aestivum* differed in the proportion 1:1.5:2 as regards chromosome length and not in the proportion of 1:2.24:3 as would be expected from the additive value of total chromosome length of the concerned genome donors. The DNA content of individual nuclei in the species studied varied in proportion to

Table 2.12.   Chromatin Length and DNA Content of Three Diploid and Polyploidy *Triticum* Species [68].

Species and Variety	Total Chromatin Length in μ	Mean DNA Content in Arbitrary Units ± SE	Ratio Chromatin Length	Ratio DNA Content
T. monococcum var. Japanese Early	127.13	2.375 ± 0.035	1	1
A. speltoides var. ligustica	157.50	2.762 ± 0.065	1.24	1.16
A. squarrosa Cambridge strain	96.88	2.219 ± 0.049	0.76	0.93
T. dicoccum var. Khapli	203.41	3.443 ± 0.044	1.5	1.45
Amphidiploid (T. aegilopoides x A. speltoides var. ligustica	240	4.05 ± 0.020	1.89	1.70
T. aestivum var. (59)	254.26	4.56 ± 0.140	2.0	1.92

chromosome length, thus indicating that the DNA content per unit length of chromosome is constant in the material studied.

The data indicate that a considerable degree of elimination of chromosomal material has taken place in tetraploid and hexaploid wheats subsequent to their origin. It is plausible that chromosome diminution has been an important factor in the conversion of tetraploid and hexaploid wheats into functional diploids. Swaminathan [68] also takes into account the famous paper of Riley and Chapman [20] and concludes, *A combination of the multivalent gene suppressor system and chromosome elimination appears to have led to a synthesis of the advantageous features of auto and allopolyploidy in tetraploid and hexaploid wheats.*

Extending the studies on the effect of linear differentiation of chromosomes on the proportionality between chromosome length and DNA content to three species of *Sorghum* characterised by differences in the number, size and stainability of chromosomes, Swaminathan and his co-workers [69] obtained the results in Table 2.13.

It is noted from the above-mentioned data that the DNA content per unit length of chromosome differs from species to species and even within the two diploid (2*n* = 10) *Para-sorghum* species, the one with a lower chromatin length has higher DNA content. At this juncture, the authors [69] bring into discussion the 'euchromatic' and the 'heterochromatic' regions of the chromosomes and the heterochromatic segments in the chromosomes have about 2–3 times more DNA and is compactly organised. In the DNA replicating S-phase (already described earlier), these segments synthesise DNA later, after the normal segments, called

Table 2.13.   Chromosome Length and DNA Content in *Sorghum* Species.

Species and Chromosome Number (2*n*)	Mean length of Chromosome Complement in $\mu \pm$ SE	Mean DNA Content Per Nucleus in Arbitrary Units	DNA Content Per μ of Chromosome
Sorghum nitidum (2*n* = 10)	61.3 ± 0.54	253	4.13
Sorghum purpureo — sericeum (2*n* = 10)	70.1 ± 0.76	241	3.44
Sorghum vulgare (2*n* = 20)	52.6	140	2.83

the euchromatic regions. Swaminathan and his co-workers [69] rightly point out that no regular relationship between DNA content and chromosome length could occur in species with varying amounts of heterochromatin. They also suggest that chromosome elimination is a common feature in polyploids and it appears from the DNA content data that the heterochromatic regions may form a large part of the eliminated segments in *S. vulgare*.

(e) Radiation-induced micro and macromutations

From both fundamental as well as applied points of analyses, Swaminathan's original contributions to the understanding of the frequency of occurrence of induced mutations with small and major effects as well as their evolutionary implications still remain as the most outstanding. There have been statements made during the 1950s and 1960s that induced mutations in their morphological and physiological effects are the same as those occurring in nature. The only effect of physical and chemical mutagens is the increase in the frequency of occurrence of the same spectrum of mutations found in nature. While this in itself is debatable, there is much more to it. Swaminathan's contributions have been absolutely magnificent since these directly address the issues which had not received the focused attention of several of his contemporary mutation breeders or, in a few cases, he has substantially added or modified the information rather vaguely assumed.

Swaminathan's masterpiece review article [70] titled, *Evaluation of the use of induced micro and macromutations in the breeding of polyploid crops* begins with the statement, *A knowledge of the various processes initiating and regulating the evolution of plant species in nature and of the means of reproducing artificially these processes is an asset which genetics has conferred upon the plant breeder of today.* Then he cites Charles Darwin who in his famous book, *The Origin of Species by Natural Selection* [71], recognised that small variations are less likely to be detrimental than large ones and endorses that for this reason, as well as because of the higher frequency of their incidence, micromutations have had a greater chance of becoming established by natural as well as human selection. Thereafter, he goes on to caution that the roles of macromutations should not be completely ignored. So, he writes, *there is, however, evidence from studies on both spontaneously*

occurring and induced mutations that drastic changes or macromuta-
tions need not always be destructive and that the role of such mutations
in the evolution of new species should not be totally overlooked.

Swaminathan [70] refers to the work of several geneticists studying mutagenesis in a wide range of organisms over the previous four decades and discusses the absence or extreme rarity of any induced mutation not already found in nature. While accepting the rarity of occurrence of such altogether 'new mutations' never ever found in nature so far, he certainly does not rule them out. Of course, these cannot be of practical interest for purposes of crop improvement by breeding in view of the rarity of occurrence of 'new' mutations.

Swaminathan elegantly explains the role of macromutations in evolution. He describes how the genus *Triticum* offers examples of the role of macromutations which at one step lead to the origin of taxonomic categories higher than varieties. For instance, the *speltoid suppressor* factor Q located on the long arm of chromosome '5A', which must have arisen as a mutation in a 'non-free threshing' tetraploid wheat, has played a dominant role in the evolution of bread wheat. The Q factor not only inhibits spelting but also brittleness of rachis, a development of obvious importance in cultivated forms. The multivalent suppressor gene located on chromosome '5B' and the Q gene located on chromosome '5A' have shaped the destiny of *Emmer* and bread wheats as staple food of a large proportion of the humans.

Swaminathan's analyses of the genic differences among the seven species in the *Triticum* group are simple and straightforward as follows:

- **Free-threshing:**
  T. aestivum
  T. sphaerococcum
  T. compactum

- **Spelt:**
  T. spelt
  T. macha
  T. vavilovi
  T. zhukouskyi

The free-threshing species viz., *T. aestivum, T. sphaerococcum* and *T. compactum* are each 'separated' from the bread wheat *T. aestivum* by a single gene: 'C' located on chromosome '2D', 'S' on '3D' and *Q* on '5A', respectively. Having presented these analyses, Swaminathan points out that a whole set of characters are controlled by each of these loci with the result that though genetically they should all be regarded as members of one species, taxonomists have described them as independent species. In a subsequent paper titled, *Macromutations and sub-specific differentiation in Triticum*, Swaminathan with his doctoral research scholar Rao [72] concludes: *In view of the fact that the key characteristics separating the 42 chromosome Triticum species are controlled only by 1 or 2 genes, all of them can be considered as subspecies of T. aestivum L. as suggested by Mac Key. A similar situation probably prevails in the tetraploid group. The important role that macromutations appear to have played in the differentiation of the tetraploid and hexaploid species of Triticum suggests that this means of variation may be a potent factor in the diversification of polyploids.*

Swaminathan has furnished several examples of macromutations in wheat, tobacco, cotton and autotetraploid barley.

The frequency of macromutations in hexaploid *Triticum* species (Table 2.14) is reproduced from Swaminathan's paper [70].

Micromutations — Swaminathan [70] and Bhatia and Swaminathan [73] describe how these micromutations are not visible to detect and

Table 2.14.   Frequency of Macromutations in Hexaploid *Triticum* Species.

Species	Number of Mutations Isolated in M2	Number of Mutations of Type		
		Speltoid	**Compactoid**	**Aestivum**
T. aestivum	206	48	16	—
T. compactum	253	04	—	77
T. sphaerococcum	219	08	02	138
T. spelta	36	—	0	0
T. macha	13	13	0	0
T. vavilovi	05	05	0	—

Source: Reproduced from Ref. [70].

require statistical methods to assess them. Two of his Ph.D. students, Dr. J.V. Goud and Dr. P.S. Bhatnagar have extensively worked on micromutations induced by physical and chemical mutagens in bread, wheat and barley. In all their studies, the radiation-induced variation for polygenically determined characters (like height, breadth of leaves, yield, etc.) has received a great deal of attention. The 'yield' of crop plants, of course, has always been the major quantitative trait. Unlike the major genes with dominant, recessive or intermediate expressions, the polygenic/quantitative genes each add a small effect towards yield and other such quantitative traits. While discussing micromutations, Swaminathan [70] refers to the radiation-induced mutation breeding in peanut (*Arachis hypogaea*) by Walton Gregory in North Carolina, USA and the yield-enhancing micromutations. Gregory [74] had obtained mutants in peanut (*A. hypogaea)* with significantly higher yields in comparison with the parent strain. Swaminathan had drawn attention to the fact that all the micromutations can be isolated and fixed only through the adoption of biometrical procedures. From the point of plant breeding for yield improvement, Swaminathan [75] refers to the work of his team that has shown the immense scope for increasing the number of ear-bearing tillers, the number of grains per ear, the weight of grains and the protein content in the grains in wheat.

In the area of micromutations induced by physical and chemical mutagens, the outstanding contributions of Swaminathan in his *The origin of macro from micromutations and factors governing the direction of micromutational changes* [76] are as follows:

(i)   The free-threshing locus Q in the genus *Triticum* has evolved through a series of tandem repeats. The basic factor $Q1$ which had earlier been identified in the hexaploid species *T. spelta* has also been found in the tetraploid *T. dicoccum* through an analysis of hydroxylamine-induced mutations. The evolutionary history of the Q locus indicates that the loci which can be regarded as having arisen through a single macromutational event may sometimes arise through micromutational steps.

(ii) So far as induced polygenic variation is concerned, the studies in varieties of *T. aestivum* and *Brassica campestris var. toria* suggest that the previous selection history greatly influences the direction of the induced variation. *There is a preponderance of mutations in a direction opposite to that of the previous selection.* He adds that if this is also true of spontaneous micromutations, the evolutionary potential of this phenomenon would be enormous. He does not find any relationship between the frequency and spectrum of chlorophyll and visible mutations and of micromutations. Therefore, it would be misleading to base recommendations on the dosages of mutagens suitable for use in mutation breeding work on the frequency of chlorophyll or visible mutations induced by a particular treatment. Fast neutrons are quite effective in creating extensive polygenic variability.

The possible origin of Q locus of various potencies as explained by Swaminathan is illustrated in Fig. 2.2.

Swaminathan refers to 'macromutations' and defines them as involving changes in a whole constellation of characters and hence having a striking phenotypic effect. A good example from his work is the isolation of non-lodging stiff-strawed mutants in wheat, rice and barley. These mutants have not only a shorter and thicker culm, but invariably also a more dense ear and more stiff leaves. Such mutants, Swaminathan points out, are of great practical interest since they confer upon the plant a morphological frame vital for it to grow under conditions of high soil fertility and abundant water supply.

Swaminathan [76] describes the 'systemic mutations' as those which either simulate an already existing taxon or necessitate the creation of a new systematic unit. He refers to his team's accomplishment in creating from one variety of bread wheat (*T. aestivum*) all the known sub-species of this species, namely *spelta, macha, vavilovi, compactum* and *sphaerococcum*. Also, from a variety of tetraploid wheat, species having key characteristics of *T. durum* and *T. pyrimidale* have been isolated. These are real good examples of artificial transmutation of one species into another. Swaminathan also refers to their isolating a mutant in the variety N.P. 797 of *T. aestivum* with adventitious branching in the ear, a characteristic recorded in nature only in *T. turgidum var.* 'mirabile' (popularly known as 'Miracle' or 'mummy') wheat. The incorporation of this branching character in commercial wheat varieties could lead to considerable gains in yield.

From the foregoing, it is evident that the mutagens-induced mutants represent a wide spectrum from the more frequent micromutations to the less frequent macromutations and systemic mutations with revelations on phylogenetic relationships. Notwithstanding all these highly impressive records, Swaminathan ends with a caveat that *mutation research should, however, be regarded only as a valuable supplement and not as a substitute for the conventional plant breeding.*

Chapter 3

Basic Research in Frontier Areas of Life Sciences

Under globalisation and patent regime, science is no longer entirely geared towards gaining knowledge for the sake of satisfying curiosity. While referring to the bygone decades, that is, 1950s and 1960s, there were water-tight compartmentalisation of botany, zoology, chemistry and physics. Even in the area of teaching 'evolution' in the botany department, no examples from animal kingdom would be appreciated. In even agricultural sciences, interaction between the departments of entomology, mycology and plant pathology was either non-existent or existent only feebly despite the fact that various insect pests are the vectors of viral diseases. Those were the times when Swaminathan broke the barriers across disciplines and provided an inter-disciplinary approach for solving several intricate problems in biological sciences. He did what none had ever done before. Agriculture as a science, to him, needed to integrate the life and physical sciences as well as the social sciences. He is a living encyclopaedia and it is gratifying that the prestigious 'Indian Agricultural Research Institute (IARI) Library' has been dedicated as 'Professor M.S. Swaminathan Library' in April 2016.

Swaminathan pioneered basic research on the effects of ionising radiation and chemical mutagens in the Botany Division of the IARI. He had procured a gamma radiation source (2000 Ci cobalt-60 gamma cell from Atomic Energy Canada Limited), a therapeutic Phillips X-ray machine and also put up a 'Gamma garden' to provide chronic irradiation to plants. As for the reason for setting up all these irradiation facilities, his answer is found in his paper [75] *Artificial Transmutation of the Gene* written for the *Journal of the IARI Post-graduate School*. He wrote, *While with some of the great practical achievements in plant breeding like the exploitation of hybrid vigour in maize, a theoretical understanding of the enhanced vigour of the hybrid has not been an essential pre-requisite for economic gains, the reverse is true of mutation breeding. The more one knows about the genetic architecture of the material and the mechanism of action of the mutagen, the greater will be the chances of getting worthwhile results. It is hence that I believe that the theoretical understanding currently being built up at the Indian Agricultural Research Institute of the various processes connected with induced mutagenesis would yield rich dividends in the near future.*

Swaminathan's ability to grasp the fundamentals of various disciplines in which he never had formal lectures and training is incredibly fantastic. In evidence, many examples could be provided. For the present, I would refer to Swaminathan's review [77] in *Current Science* of the book entitled *Mechanisms in Radiobiology* edited by Errera and Forssberg [78]. Even after 50 years, a pertinent question remains unanswered as to how a crop cytogeneticist from IARI, New Delhi rather than a hard-core radiobiologist from Bhabha Atomic Research Centre (BARC), Bombay was entrusted with the task of reviewing a book on *Mechanism in Radiobiology*. The review written by him in 1962 for the book edited in 1961 is still valid even after five decades. The review makes a critical review of the dosimetry, ionic yields in gases, water and organic liquids, the steps in the sequence of development of radiobiological damage initiated with the absorption of the photon energy in the cells to striking morphological/ physiological effects, and above all, Alper's altogether different and extremely revolutionary ideas on the importance of targets other than the DNA in causing radiobiological damage. Alper's papers implicating cell membranes rather than the DNA as the major target for the ionising radiation ought to have been highly unacceptable and even repulsive to most radiation geneticists, but Swaminathan deals with Ticvah Alper's views with respect and scientific curiosity.

At this point, I would like to share a personal anecdote. In 1993, I was invited to be the Director of the Bioscience Group of the BARC, Mumbai.

Shifting from Jawaharlal Nehru University (JNU), New Delhi did not seem to be a bright idea, but I did after having been told that I would be the first ever, and also possibly last ever person to be directly appointed to a highly responsible position at BARC from outside of the Department of Atomic Energy (DAE) system. At BARC, I gave a lecture on unsolved problems in radiation biology, and how possibly to deal with them. Dr. P.K. Iyengar, a former Chairman of the Atomic Energy Commission and then President of the Indian Nuclear Society chaired my lecture. At the end of the interaction with some of the top scientists (physicists, chemists, radiation chemists, radiation biologists, nuclear medicine experts and so on), Dr. P.K. Iyengar greatly appreciated the lecture and asked me as to where I had initial education and training in radiation biology. I replied that it was from Dr. M.S. Swaminathan at the IARI, New Delhi. Dr. P.K. Iyengar was totally dumbfounded for a while, and then after a pause, he said that he knew Dr. M.S. Swaminathan only as an agricultural scientist *par excellence*. This anecdote might provide to the readers some idea of the great capacity of Swaminathan to master vastly different areas with such ease and elegance.

3.1 Swaminathan and Basic Radiobiology

Swaminathan's basic research in radiation biology included elucidation of phenomena such as *oxygen effect* in relation to the linear energy transfer (LET) of the radiations, relative biological effectiveness (RBE) and more fundamentally what was then beginning to be known as the 'direct' and 'indirect' actions/effects of ionising radiation. Before discussing some of his original research papers in the areas mentioned above, I, a beneficiary of his most inspiring lectures in 'Radiation Genetics', would like to make a brief mention of him as a teacher. For the post-graduate students of the Post-graduate School of the IARI, he taught a basic course, the cytogenetics-I, and 'Radiation Genetics', an advanced course. Without any doubt, he was indeed the most inspiring teacher. He would ride on his bicycle from his Bungalow No.12, via experimental field to the class room, always a few minutes before time (8 am). His adherence to punctuality was so very compelling that some of the students waking up late in the morning would have to skip breakfast in order to reach the class room in time. To Swaminathan, punctuality was non-negotiable, and he was always on time, no matter whether rain or shine, summer or winter. Once one listened to his inspiring lectures, there was hardly any need for reading a textbook. Of course, one can go for the original research papers, since he always provided a list of references. In 'Radiation Genetics', he covered basic mechanisms — the formation of free radicals following the absorption of radiant energy by cells and organisms, the photoelectric effect, the Compton effect, the pair production following irradiation with X-rays, gamma-rays, the reaction of free radicals with oxygen, or with cellular organelles and macromolecules, DNA damage by direct and indirect actions of ionising radiation, genetic and cytogenetic consequences, role of molecular oxygen in bringing about differential radiosensitivity between normal and hypoxic cancer tumour cells and also therefore the relative advantages of using neutrons for cancer radiotherapy but for the cost factor and convenience of gamma-rays, etc. He would refer to the work of 'Radiation Effects Research Foundation (RERF)' in Japan and enlighten the students with current trends of research of top American and Japanese geneticists on the possible genetic effects of atomic bombs in the survivors and their children.

Swaminathan's lectures in an ever-advancing field of Radiation Biology were more based on results of contemporary research; one essentially needed to keep abreast of new developments through access to original publications in journals such as *International Journal of Radiation Biology* (UK), *Radiation Research* (USA), *Radiation Botany* (USA) and a few others. He had facilitated the IARI library to subscribe for these journals. Swaminathan was also a member of the editorial board of *Radiation Botany* (Pergamon Press, New York).

Personal contacts with global leaders in science (surely Swaminathan was already internationally well known and highly respected) greatly helps the developing science in India. During 1950s and 1970s, Swaminathan was a close friend of several top leaders in radiation biology, the Nobel Laureate H.J. Muller, A.H. Sparrow, Lars Ehrenberg, M.L. Randolph, Ake Gustafsson, Sheldon Wolff, Gunnar Ahnstrom, R.A. Nilan, Ralph Singleton, Arne Muntzing, C.F. Konzak, T.S. Osborne, F.D. Amato, Von Wettstein, W. Gregory, R.D. Brock, H. Gaul, G.T. Scarascia, H. Glubrecht, A. Suss and several others from USA, UK, Europe and

Japan. Yet, the truth is that friendship among scientists working in a particular field of science is established through hard work and original research papers published in the reputed international journals. With this in mind, I have enlisted a few papers in addition to those already cited with reference to polyploidy and radiosensitivity. Some of these are described below.

Since 1926, the prevalent hypothesis of radiation action on cells was purely physics-based, that is, the energetic photons 'hit' the vital target molecule in a cell like a 'bullet' and cause its inactivation/death. That was the 'target theory'. This target theory ignored some of the observations, reported a decade earlier, such as (i) rapidly dividing cells are more radiosensitive, and (ii) that oxygen enhances the radiobiological damage induced by X-rays. A major challenge questioning the *target theory* came in 1947 from two British radiobiologists Thoday and Read [79], who published a paper, *Effect of oxygen on the frequency of chromosome aberrations produced by X-rays*. Soon several other papers emphasising the role of oxygen in enhancing radiobiological effect induced by low LET ionising radiations (250 kVp X-rays, gamma-rays from cobalt-60, cesium-137). Interestingly, Swaminathan was among the first and very few to point out that neutron radiation with higher LET and higher RBE does not have a discernible oxygen effect (i.e. almost nil) and therefore, it would be advantageous to use neutrons than X-rays or gamma-rays in cancer radiotherapy.

In his paper [80], *Use of Neutron Irradiation in Agriculture and Applied Genetics*, jointly written with Dr. A.R. Gopal-Ayengar, then Director of the Bio-Medical Group of Atomic Energy Establishment Trombay, Bombay, Swaminathan shows (see Table 1, Ref. [80]), that the influence of oxygen as a modifier of radiobiological damage is 'appreciable' for X-rays (and gamma-rays) and 'none' (or negligible) for neutrons (Table 3.1).

Continuing the discussion on radiobiological mechanisms, the 'target theory' lost its tenability particularly after the unequivocal demonstration that the first step in the development of the radiobiological effect in the radiolysis of water in the cell is generation of free radicals such as hydroxyl radicals, hydrated electron, singlet oxygen, etc., and the formation of a 'reactive oxygen species' (ROS). The question, however, was whether these reactive entities resulting from radiolysis of water could

Table 3.1.   Biological Effects of X-Rays and Neutrons [80].

Type of Observation	Value of Reaction for	
	X-Rays	**Neutrons**
(a) Dose necessary to produce any biological effect (in barley)	Great	Small
(b) Influence of modifiers: O_2, N_2, H_2S	Appreciable	None (or small)
(c) Influence of modifiers: water content, irradiation temperature, narcosts	Appreciable	None (or not investigated)
(d) Effects of pre-soaking	Great	Small, but appreciable
(e) Effect of liquid air temperature on irradiation	Appreciable	Insignificant
(f) Effect of chromosome number on radiation induced growth inhibition	Rel, small	Rel, great
(g) Effects of properties of individual cells ('Intrasomatic selection') or seeds	Great	Small, or, at least in some cases, none
(h) Ontogenetic development of injury	Plants die soon after irradiation or recover and continue growth	Plants die when mitoses become of importance for growth, without recovery
(i) Development of chlorophyll deficient regions	Necrotic, turgorless cells are formed, which die or recover	Destroyed plastids in otherwise healthy, turgescent cells

cause biological damage. That these reactive species indeed induce biological damage was shown by Stone *et al.* [81] in their paper, *The production of mutations in Staphylococcus aureus by irradiated substrata.* In a subsequent paper, Stone [82] noted mutations in organisms 'not' directly exposed to ionising radiation, but were cultured in the irradiated substrata. He referred to such effects of ionising radiation as *indirect effects* of radiation. That indirect effects of radiation arising from hydroxyl radicals ($^{\cdot}OH$); hydrogen radicals ($^{\cdot}H$) and hydrated electrons (e^-aq) break the chromosomes of diploid, tetraploid and hexaploid wheat was shown by Natarajan and Swaminathan [83]. They allowed the seeds of the three wheat species to germinate in irradiated (X-rays ranging from 6,600 R to 1,32,000 R from a Phillips X-ray equipment) water and fixed the root tips as these emerged after 48 h. They also cultured embryos from germinating seeds of *Triticum monococcum* and *T. dicoccum* in White's medium irradiated with 33,000 R of X-rays.

The data on chromosome aberrations showed that while chromosome stickiness was the only abnormal feature observed in root tip cells of embryos cultured in normal White's medium, chromosome and chromatid breaks were observed in those cultured in irradiated medium.

The radiation chemistry of direct and indirect effects of ionising radiation is briefly as follows:

Gamma/X-rays

1. Water (H_2O) $\wedge\!\!\!\vee\longrightarrow$ ·OH (hydroxyl radicals); e⁻aq (hydrated electrons); ·H (hydrogen radicals)

2. When oxygen is available during irradiation of water (oxygen solubility in water is greater at about 10°C than at higher temperatures), the products formed are as follows:

   - e⁻aq + O_2 → O_2·⁻ (superoxide anion)
   - ·H + O_2 → ·HO_2 (hydroperoxide)

   These react with DNA and other vital molecules in the cell and damage them. The hydroxyl radical (·OH) can attack DNA molecule and abstract electron from it as follows:

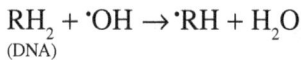

$$RH_2 + \text{·OH} \rightarrow \text{·RH} + H_2O$$
(DNA)

The ·RH is a potential lesion in the DNA. When it reacts with oxygen in the cell, it results in irreparable damage:

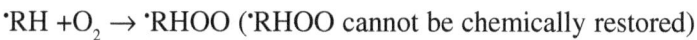

$$\text{·RH} + O_2 \rightarrow \text{·RHOO} \; (\text{·RHOO cannot be chemically restored})$$

In the absence of oxygen, the ·RH can be chemically restored as follows:

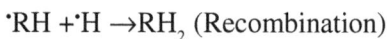

$$\text{·RH} + \text{·H} \rightarrow RH_2 \; (\text{Recombination})$$

The above modes of ionising radiation are referred to as 'indirect action' of radiations. On the other hand, the direct action of radiation on DNA is depicted as follows:

$$RH_2 \xrightarrow{\text{Gamma-ray}} \,{}^{\cdot}RH + {}^{\cdot}H$$

When oxygen is present, the potential lesion (·RH) forms ·RHOO (irreparable damage) as shown in the case of indirect effects. Based on all the information on the radiolytic products of water, and their deleterious actions on the DNA (i.e. chromosome breakage), Swaminathan and his co-workers published a series of papers in top journals such as *Science, Radiation Botany, Die Naturwissenschaften* and *Radiation Research* bearing on the implications of the indirect effects of radiation in the radiation preservation of readily perishable foodstuffs, radiation disinfestation of food grains under storage conditions, and radiation-induced delay in sprouting of onions, potatoes, garlic, etc. In one interesting study, Swaminathan and his co-workers [84] published a paper which showed that an increase in the mutation rate occurred in *Drosophila melanogaster* (fruitfly) reared on a basic medium irradiated with sterilising dose (1,50,000 R) of cobalt-60 gamma-rays. In Muller-5 tests, sex-linked recessive lethal occurred only in the F2 progenies of the male test flies obtained from breeding the parent flies on irradiated medium, while visible changes occurred in experimental cultures from both the control and irradiated media. The frequencies of sex-linked recessive lethal were 0.35%, 0.55% and 0.8% in three independent experiments. Swaminathan and his co-workers [85] reported chromosomal breaks in root tip cells of onion *(Allium cepa)* and barley (*Hordeum vulgare*) cultured on irradiated fruit juices. These juices were irradiated with 2,00,000 R X-rays. Sprout inhibition and radiomimetic properties of irradiated potatoes were also demonstrated [86]. Male flies (*D. melanogaster*) fed on irradiated medium showed further increase in crossing-over [87]. Swaminathan [88] led the team of researchers on the indirect effects of ionising radiation with the first paper in the series by reporting cytological aberrations in barley embryos cultured on irradiated potato mash.

In 1963, I, after my selection for admission to the Post-graduate School of the IARI, New Delhi, was fortunate to join the doctoral research programme under the guidance of Professor M.S. Swaminathan then, Head of the Division of Botany. Swaminathan had either an intuition or an inimitable analytical capacity to quickly assess his students' potential for diligence, aptitude, intellectual potential (and 'what else' I would not know) in assigning research problems to them. In the PG course at the Post-graduate School of the IARI, course work was compulsory. There was also a system of Comprehensive Examination which a post-graduate student should pass before he/she would be free to devote full time for his laboratory/field studies. No student to the best of my knowledge had ever declined to work in the research problem assigned to her/him by Swaminathan.

When I joined Professor Swaminathan, he suggested cytogenetics as the major course work and genetics and physics for the two minor courses. The minor in physics consisted of atomic physics, nuclear physics and I really had to put in very hard work to perform satisfactorily in these courses.

The reason for asking me to do physics was that my research on the 'direct and indirect actions of ionising radiation' was to elucidate the mechanisms and not use radiation merely as a tool to induce mutations in crop plants. By late September 1963, it became evident to Swaminathan that I had done well in the major and minor course work. He then put me on whole time research.

Swaminathan's Laboratory Number 4 (Lab 4) was the central place where he would regularly meet his students over a cup of coffee at 11 am. The students would prepare the coffee on a rotation basis. Each one of the students as well as Professor Swaminathan had a personal coffee mug, and after finishing the coffee over scientific discussion, each one (including Swaminathan) would wash his/her coffee mug and put it in its assigned place. Sometimes the discussions would be in a lighter vein, and everyone would get a dose of 'laughter' as medicine. Yet, at other times, it would be more serious and the discussion could be on the basis of work done. So, it was on one such regular session, Swaminathan entered Lab 4 with a piece of hand-written paper with the title, *Indirect Effects of Radiation*. This title had a few sub-titles and one of these was *oxygen effect*. At that time, I had not credited the course 'Radiation Genetics' taught by Swaminathan. So, I did not have even the slightest clue regarding *oxygen effect*. So, I went to the well-equipped IARI library (now Professor M.S. Swaminathan Library) and picked up a couple of textbooks on the *Actions of Ionising Radiation on Living Cells.* One of the best textbooks in Radiation Biology in those days was the one written by Douglas Lea (Cambridge University, 1950). The book was about 400 pages, and it had mentioned the word *oxygen* in a couple of places, but nothing much of its effects (*oxygen effect*) on cells and organisms exposed to ionising radiation. It was much later that I realised that Swaminathan had read the paper [79] *Effect of oxygen on the frequency of chromosome aberrations produced by X-rays* by Thoday and Read. In the following trimester, I credited Swaminathan's 2-credit course on 'Radiation Genetics' that was so inspiring and even after five decades, I do at times hear in my dreams Swaminathan's lectures on *oxygen enhancement* of biological effects of low LET ionising radiation, formation of free radicals, direct and indirect effects of radiation, chemical radioprotectors and radiosensitisers, RBE as a function of the LET, radiosensitivity modified by physical factors, what could possibly be expected from the work of RERF, Japan which was studying the genetic effects of the atomic bombs on the survivors and their descendants, radiation mutagenesis for crop improvement, diplontic selection, etc. I and surely lots of others would love to relive those days.

So, I started my research on indirect effects of ionising radiation based on the strong foundation already laid by Swaminathan and his team of researchers. I used human lymphocytes *in vitro, D. melanogaster, Tropaeolum majus* and barley as the test systems. I published my research findings jointly with Professor Swaminathan in *Radiation Botany* [89, 90], *Indian Journal of Genetics* [91] and *Current Science* [92]. Our results clearly revealed that solutions rich in sugar (sucrose) undergo radiolysis to form aldehydes and ketones; the pH of the solution is lowered with increasing dose of radiation. These cause the observed *radiomimetic* effects. However, these effects rapidly disappear with storage especially at higher (25–35°C) temperatures which indeed is the major objective of radiation disinfestation of the grains during storage; the authors published a critical review paper taking their own as well as data of other published papers. This review paper was published in *Radiation Botany* [93].

The review paper made a few conclusions that have implications for the use of radiation technology for extension of shelf-life of a variety of food materials. First, the radiation source should be gamma-rays emitted by cobalt-60 or cesium-137. Alternatively, electron beam of up to 10

MeV could be used. Second, milk, fruit juices and food materials rich in sugar and water are not suitable for radiation preservation. Third, radiation sterilisation of beef, pork and other such meats can be done using radiation doses of up to 10 kGy. Above 10 kGy, the palatability is adversely affected due to cross-linkages of proteins. Food grains can be disinfected at moderate doses (75 kr or 0.75 kGy). Immediately, after irradiation, the dry seeds (up to 7% moisture content) have been shown to have the radiation-induced free radicals, which decay with the duration of storage and also more rapidly at higher temperature. Higher temperatures facilitate radical–radical recombination. Higher water content (>9.0–10.0%) of the seeds also facilitate harmless recombinations of the radicals.

Very small doses of ionising radiations could be used to inhibit sprouting of onions, potatoes, garlic, etc. In the late 1960s, the BARC at Mumbai, Government of India had sought the permission of the Union Ministry of Health and Family Welfare, Government of India to commercialise food irradiation for purposes such as the sprout-inhibition of onions, potatoes, shallots, garlic, etc., for disinfestation of wheat and a few other grains before their storage and radiation sterilisation of pork, beef, chicken, fish, etc., for extension of their shelf-life. It is then that the National Institute of Nutrition (NIN), Hyderabad reported findings of cytogenetic abnormalities (especially the occurrence of polyploidy cells) in rats, monkeys and children fed *chappatis* (or other preparations) from wheat exposed to 75 kr. The BARC in Mumbai, on the other hand, had published numerous papers in leading international journals that products for human consumption made from the irradiated wheat did not cause any cytogenetic abnormalities and particularly the polyploid cells in rats and monkeys. The BARC, however, did not include children in their studies on ethical considerations. Conflicting data on the clastogenic and polyploidising effects of irradiated (75 kr) wheat coming from two great national institutions urgently needed to be resolved. The Government of India identified Swaminathan, then Director-General of the Indian Council of Agricultural Research (ICAR), New Delhi as the eminently suited person and requested him to resolve the issue. After chairing a few meetings with the top officials of the NIN and BARC, Swaminathan decided that it was necessary to reanalyse the raw data from both the institutions as well as their experimental designs. He also noted that the causes of conflict fall into both biological and statistical realms. He therefore recommended me, then an Associate Professor in the School of Life Sciences of the JNU, New Delhi and Professor P.V. Sukhatme, an outstanding authority on statistical methods and analyses of those days, to the Government of India to form a Committee. So, it was formed. Their painstaking analyses revealed that the NIN was yet to develop the requisite expertise in the design and execution of cytogenetic studies in the mammalian systems, and that there were also serious flaws in the statistical procedures. The reference to the conflict between NIN and BARC is purely incidental to what is intended to be conveyed in this biography. The notable point is food safety is non-negotiable to Swaminathan. He would insist that the science on which a technology is based is flawless and sound. In this particular case, he first led a team to understand the indirect effects of radiation, then he designed appropriate experiments to demonstrate the radiomimetic effects of irradiated substrata on the unirriadiated organisms fed/cultured on them and finally assigned a doctoral research programme to one of his students (me) with a view to develop human resource adequately trained and competent to tackle problems in taking a new technology to the society and the nation. What Swaminathan had elegantly done for safety evaluation of foods treated with radiation technology has not, unfortunately, so far been done for genetically engineered (i.e. genetically modified) foods.

3.2 Top-Down from Meiosis of Potato and Wheat to Mitosis of Yeasts

Earlier chapters have brought out Swaminathan's outstanding contributions to the study of chromosomes in mitosis and meiosis of potato and then wheat and how based on these studies, he has elucidated several aspects of their phylogeny, intra- and inter-specific variations. In this section, the focus is on Swaminathan and his student Ganesan [94] unambiguously demonstrating that lower eukaryote yeasts (*Saccharomyces cerevisiae Zygosaccharomyces priorianus and Schizosaccharomyces pombe*) also have a well-defined nucleus as also the stages (interphase, metaphase, anaphase and telophase) of mitosis as in higher eukaryotes. When Swaminathan and Ganesan initiated their work in yeast around 1956–1957, there were many wild guesses such as that the nucleus in yeast forms an 'endopolyploid complex' and it is intra-vacuolar. Yet another view was that this is a binucleate system, with the vacuole being the vegetative macronucleus, and there would be a smaller regenerative nucleus as in the ciliate organisms (e.g. *Paramaecia*). The old textbooks have described the nuclear fission in yeast as 'amitosis'. The reason for the lack of clarity on the nuclear division in yeast is largely due to lack of specific staining procedure to achieve a clear picture of the yeast nucleus and its division. Therefore, as the first step Ganesan and Swaminathan [94] developed a somewhat long and tedious technique to expose the yeast nucleus embedded within a complex network of nuclear membrane and then stain it with Giemsa. Their technique introduced several salient steps. The first one was to keep the yeast cells young by repeated sub-culturing. These young yeast cells were centrifuged, washed and smeared on slides coated with a little egg albumen. These cells were treated with 0.002 M 8-hydroxyquinoline for 1 h, fixed first in osmium tetroxide (O_sO_4) vapour for 30 s and then in chloroform for 30 s. The slides were passed through descending grades of alcohol, washed in distilled water, immersed in 0.17 M NaCl solution for 2 h at 57°C. They were then washed in distilled water and later hydrolysed in 1N HCl at 60°C for 5–7 min. This was followed by washing in distilled water and buffer. The slides were kept for 3 h in Giemsa stain comprising 96 mL of phosphate buffer of pH 7.0 and 4 mL of the stain. After dehydration, mounting was done in balsam. Nuclei were brightly stained and well differentiated; centromeres were clear and the process of nuclear division and movement to the daughter cells could be studied. Pretreatment with 8-hydroxyquinoline increased the viscosity of the cytoplasm, while NaCl treatment and acid hydrolysis led to complete removal of ribonucleic acid and basophilic material. A selective staining of chromatin was thus achieved. Structures resembling chromosomes could be seen when fixed and stained cells were squashed, soon after the removal of the slides from the stain, under a cover glass by applying uniform pressure with a rubber stopper.

In a biography, the materials and methods of an experiment are rarely described; but in the case of Swaminathan's work on yeast mitosis, careful choice of various chemical substances as well as duration of the exposure of the 'young' yeast cells in sequence with each of them have been the key to the success. After first demonstrating the occurrence of typical mitosis in yeast cells, Swaminathan and Ganesan [95] proceeded to assess the kinetics of mitosis in these cells. With the proof of excellent photographs of yeast cells in division, the authors stated:

Division phase — at the onset of mitosis the nucleus is resolved into chromosomes, and configurations resembling the prophase and meta-phase chromosomes of other organisms could be seen in squash prepara-tions. The exact chromosome number is, however, not clear. We have seen eight chromosomes in some, but not in all cells. The difficulty in getting consistent picture of chromosomes arises largely from the persistence of the nuclear membrane. Metaphase is followed by anaphase, in which the two chromatids of each chromosome seem to go to the opposite poles; telophase then sets in. We have never observed directly anaphase and telophase within the nuclear membrane, but the pattern of staining and appearance of unstained regions within the nucleus suggest that these stages follow metaphase.

The authors have elaborated the *separation phase*. The centrosome divides into two at the end of the intra-nuclear mitosis. Dismissing that yeast nuclei divide by budding, the authors stated: *The doubling was observed in many yeast nuclei at the end of mitosis but before separation of the daughter nuclei, as well as the lack of any relationship between budding and nuclear separation either in time or in plane of movement, supports our view that anaphase and telophase are actually completed within the nuclear membrane, thus effecting a regular distribution of the divided chromosomes prior to the initiation of a process analogous to cytokinesis in the nucleus.*

Another notable paper by Swaminathan and Ganesan [96] *Mitosis in Yeasts* was an exceptionally befitting contribution to the *Memoirs of the Indian Botanical Society*. Their research confirms that yeast is a eukaryote with nuclear division, mitosis.

3.3 Pusa Institute's *Gamma garden*

A note of the radiation source

Normally, since the days of Stadler [47], the plant materials intended for irradiation with either the X-rays or gamma-rays from cobalt-60 or caesium-137 had to be brought into the rooms or small chambers where the irradiation sources are installed. These are specially shielded for drastically reducing the background radiation doses in order to ensure maximum safety of the workers and researchers outside of the radiation facility. As one could readily visualise, it used to be, and still is, cumbersome. Further, it poses a limitation of the volume of the plant material that could be exposed; not only that, even more difficult is to expose reasonable quantities of the plant material at different doses and dose-rates. The X-ray and gamma-ray irradiators fixed in the rooms have a highly limited range of control of the dose-rates. Under these circumstances, one can readily appreciate quite a high level of flexibility that could be achieved when the irradiation source (say a cobalt-60 gamma-rays) is located in a deep pool of water and the source is raised from its pool of water using a remote electronically operated device to irradiate the plant materials kept at varying distances from the source. This facility is shielded with thick concrete wall with about 10–12 ft of height. The plants would receive doses at dose-rates decreasing with increasing distance from the cobalt-60 source. This facility described above was actually set up by Swaminathan at the IARI, New Delhi on August 25, 1960. This was called the *Gamma garden* of the IARI, New Delhi and is described here:

- ***Gamma garden* at the World Agricultural Fair, New Delhi, 1959–1960:** The world's first *Gamma garden* was set up in early 1950s at the centre of an isolated area in Long Island near New York. Because of the timely initiative of Swaminathan, India was not far too behind in setting up a *Gamma garden* just 10 years later. The genesis of the Indian *Gamma garden* at the IARI was influenced by the design and use of a cobalt-60 Irradiation Unit in the United States Exhibit, World Agricultural Fair in New Delhi from December 11, 1959 to March 01, 1960. The unit was designed for the gamma irradiation of biological and other research material as well as for demonstrational purposes and operated as a service unit to scientists throughout the fair. The design of the facility with 4,750 curies of cobalt-60 as the source of gamma-rays has been described by Brewbaker and Swaminathan in *Current Science* [97]. The readers are recommended to read the above-mentioned paper to get the details of the structural design of the Gamma Irradiation facility at the World Agricultural Fair 1959–1960 in New Delhi.

- *Gamma garden* **set up at the IARI, 1960 (chronic irradiation facility):** Impressed with immense opportunities to irradiate a variety of crops at different dosages and dose-rates, Swaminathan mobilised resources and administrative support to set up the *Gamma garden* at the IARI. The World Agricultural Fair ended in March 1960 and the *Gamma garden* of the IARI was inaugurated by then Union Minister for Agriculture S.K. Patil in August 1960. By any yardstick of assessment, the credit should go to Swaminathan to have set up the IARI *Gamma garden* within about 5 months after the World Agricultural Fair ended. The IARI Gamma Garden is shown in Fig. 3.1.

Figure 3.1. *Gamma garden* set up at the IARI, New Delhi.

However, some details are provided here for the readers to appreciate the safety aspects first and then the irradiation procedure. The *Gamma garden* consisted of an area of three acres, and the entire area had been encircled by a wall, 3 f thick and 12 f high, built with bricks on either side and earth compacting in the middle. The wall served two purposes, first, it offered a protective shielding so that no harm could be done to the health of the individuals working at the outskirts of the garden, and second, it helped regulate entry and work in the garden. Outside the brick wall, a barbed-wire fencing had been erected to ensure that no unauthorised person could enter the garden. The radioactive cobalt-60 source was in the form of small pellets weighing about 6 g. These pellets were inside an aluminium capsule and the capsule is welded to the lid of the lead container in which the radioactive cobalt was kept. The strength of the source was 200 curies. The radioactive cobalt source was raised from the lead container by lifting the lid of the container. The source can be raised up to 5 ft from the ground, and as soon as it is taken out of the lead container, the whole garden receives gamma radiation. Plants were grown in concentric circles and the whole garden was divided into eight sectors, each sector being allotted to a specific group of plants such as cereals, pulses, fibre crops, vegetables, fruit trees, etc. Irrigation water was pumped through hydrants fixed at regular intervals. Two radiation monitors, one fixed to the circular wall and another which was portable, could be moved radially to measure accurately the radiation doses received by the plants. Naturally, the plants kept near the radiation source would receive much higher doses and those farther away much less. Any desired dosage could be given by manipulating the duration of the treatment and the distance from the source. Thus, the IARI *Gamma garden* was a versatile irradiation unit in which a wide range of plant material could be exposed to radiations at any desired stage (i.e. seedlings, plants with initiation of flowering, seeds, etc.) in their life-cycle. The goal was not just plant breeding. The *Gamma garden* was used to induce mutations in fungi and beneficial insects, control of insect pests through the *male-sterile technique* using the sterilising properties of radiation and also to assess the usefulness of radiation preservation of readily perishable food materials. Swaminathan had further considered the problem of *diplontic selection* i.e. in which the fewer number of cells with possible mutations induced by irradiation are overwhelmed by the larger number of normal cells (i.e. not mutated by the irradiation) and consequently, the mutated cells do not multiply as rapidly as the normal cells, and hence get eliminated. However, the *diplontic selection* could be neutralised when the radiation dose was administered chronically at very low doses over a protracted period of time. In a later part of this chapter, the consequences of *diplontic selection* induced in diploid and polyploidy seeds following the absorption of radiant energy are discussed. The point to be noted is that one acute dose generally induces *diplontic selection* — a process in which a few of the mutated primordial cells are at a disadvantage in propagation in comparison with the unmutated/unaffected cells. However, with low-dose chronic exposure, more and more cells are progressively affected, and the rigour of the *intra-somatic competition* (diplontic selection) is drastically reduced. In the following section, further discussion on this subject is provided.

- **The *Diplontic Selection* in Diploids and Polyploids following seed irradiation**

It was R. Bauer who first used the term *Diplontic selection* in the context of the induction of vegetative mutations in *Ribes nigrum* [98]. An understanding of the role of diplontic selection in rendering mutation breeding efforts, rather limited in success in vegetatively propagated plants, led to the development and standardisation of techniques which would enable cells carrying mutations to express themselves phenotypically. Since radiant energy is absorbed by the multicellular systems in a manner of 'Poissonian distribution', not every single cell absorbs the radiation dose equally; in fact, several cells in a tissue might not even receive any dose at all. This explains the prevalence in seeds with a very few primordial (meristematic) cells with a greater chance of resulting in the M1 plants expressing high frequencies of mutations than in the M2 population.

As an outstanding analyst and an original thinker, Swaminathan extended the analysis of the problem of diplontic selection to yet another situation. He applied the concept of *diplontic selection* or the *intra-somatic selection* in the seed-propagated plants, that are true diploids [99]. He noted that in seed propagated plants problems arising from diplontic selection were relatively more acute in diploids than in polyploids, since in the latter, cells with intra- and inter-genic mutations tend to be viable and to be at no great competitive disadvantage owing to the presence of duplications at the genic as well as chromosomal level in the constituent genomes. He also pointed out that the extent to which this buffering effect is applicable, however, would likely vary very greatly in different polyploid species. With a view to study this problem in depth, Swaminathan [99] designed appropriate experiments involving Upland cotton (*Gossypium hirsutum* $2n = 4x = 52$) which is an allotetraploid capable of tolerating 'only' segmental deficiencies in their chromosomes, and bread wheat (*Triticum aestivum* $2n = 6x = 42$), an allohexaploid which has been known to tolerate *nullisomics* ($2n = 40$, because of the loss of one homologous pair of chromosomes) and genomic substitution.

Following Chromosomal Exchanges/Aberrations in diploid ($2n = 26$) and Tetraploid ($2n = 52$) Cotton Species [99].

(a) M1 generation

Material	Number of Chromosomal Exchange/Aberrations
Seeds of *G. arboreum* ($2n = 26$) and *G. hirsutum* ($2n = 52$) were exposed to 36,000 R of X-rays	
G. arboreum	1 in 20 M1 plants
G. hirsutum	9 in 20 M1 plants

(b) M2 and M3 generations
No phenotypically visible mutations occurred in the M2 and M3 progenies of both *G. arboreum* and *G. hirsutum*

(c) Recurrent Irradiation
Seeds of M1 plants of *G. arboreum* and *G. hirsutum* treated with 36,000 R of X-rays were treated again with the same dose of X-rays during the following year. The data indicated that the recurrent irradiation does not affect survival in *G. arboreum* more than the initial dose, but lethality increases with repeated irradiation in *G. hirsutum*. There was an accumulation of lethal as well as visible mutations in the population of *G. hirsutum* subjected to a second generation X-radiation treatment. In *G. arboreum*, however, recurrent irradiation had no effect on mutation frequency (Table 3.2).

Table 3.2.   Effect of Recurrent Irradiation on Mutation Frequency in Diploid and Tetraploid Cotton [99].

Species	Number of Generations of Irradiation	Number of M2 Families Raised	Number of Families Segregating for Mutation
G. arboreum	1	49	0
G. arboreum	2	56	0
G. hirsutum	1	80	0
G. hirsutum	2	72	6

- **Findings in hexaploid wheats**

With reference to the effects of seed irradiation on the yield of mutations in hexaploid species of *Triticum vis-à-vis* the diploid and tetraploid cotton species, Swaminathan [99] makes the following important statements: *While in the ontogenic development of cotton the first evidence of a fruiting branch is a primordium in the axil of a leaf on the main stem at the second or third node from the meristem, 3–4 independent ear primordial are already differentiated in the seeds of Triticum species, and each ear appears to develop from one initial cell. This feature, together with the ability of polyploid Triticum species to tolerate whole chromosome deficiencies, renders the occurrence of phenotypic variation probable even in the M1 generation.* Chimeras occur frequently in the M1 generation and screening of the M2 progenies for mutations bears out Gustafsson's statement that in bread wheat *with suitable X-ray doses, a mass mutating sets in.* He pointed out that visible mutations occur in a polyploid only at loci in which phenotypic buffering induced by duplications do not exist. With experimental data he stated, *A comparative study of the induced mutation rate in the six commonly recognised hexaploid species revealed that the species which lack the Q factor (namely T. spelta, T. macha and T. vavilovi) had a significantly lower mutation frequency that as compared to T. aestivum, T. compactum and T. sphaerococcum which also possess the Q factor* (Table 3.3). It has already been discussed earlier that *T. spelta, T. sphaerococcum* and *T. compactum* are each separated from *T. aestivum* by a single gene *Q* located on chromosome IX (5A), 'S' on XVI (3D) and 'C' on XX (2D), respectively.

Table 3.3. Influence of Q Factor on the Induced Mutation Rate in Hexaploid *Triticum* Species.

Species	Q factor	Number of Mutations Per M2 Family
T. aestivum	yes	60.95
T. compactum	yes	75.07
T. sphaerococcum	yes	65.77
T. spelta	No	10.40
T. macha	No	6.84
T. vavilovi	No	3.40

Source: Table II of Ref. [99, p. 284].

I have modified Table II in [99] so as to make it shorter and to focus only on the Q factor in relation to the frequency of induced mutations [Table 3.3].

With the support of enormously high quality data, Swaminathan summarised: *From studies in diploid and tetraploid species of cotton and in six hexaploid species of wheat, it was found that the nature of polyploidy as well as the histo-genetic aspects relating to the origin of inflorescences condition the mutation rate observed in them following seed irradiation. Recurrent irradiation of seeds helps to overcome to some extent the difficulty of finding visible mutations in the progenies of tetraploid cotton irradiated once. The effect of diplontic selection is not as drastic in wheat as in cotton, since 3–4 independent ear primordial are already differentiated in the seed subjected to irradiation and each ear appears to develop from one initial cell. The spectrum of induced mutation is narrow in bread wheat and serves as a measure of the extent of gene duplication existing among the different genomes. Among the loci at which there is no buffering effect and visible mutations are possible, some are relatively more radiosensitive and readily mutable.*

Swaminathan's statement in 1961 that mutation breeding offers considerable scope for incorporating a desirable additional attribute in a highly bred strain of wheat has since been proven beyond any doubt. This statement of Swaminathan needs to be viewed in the context of earlier mutation breeders who believed that mutation breeding would be of highly limited success in the hexaploid wheat.

- **An unparalleled Leader in Mutagenesis and Mutation Breeding: Excelling in experimental muddling through**

The year was 1968 — the Yield Revolution in wheat had been accomplished in the previous year and Swaminathan had already introduced chemical mutagens alongside the physical mutagens. Using chemical mutagens such as the monofunctional alkylating derivative of Acridine, Nitrosoguanidine, 5-fluorodeoxyuridine (FUDR), Hydroxylamine (HA), Ethyl-methane-sulphonate (EMS), Swaminathan and his students studied their clastogenic and mutagenic activities in *D. melanogaster, Vicia faba*, in *Emmer* wheat, barley and rice. In fact, Swaminathan and Natarajan [100] demonstrated the chromosome breaking activity (clastogenic) of vegetable oils and edible fats. A year later, they [101] reported the chromosome spreading induced by vegetable oils. An important paper by Swaminathan and Natarajan [102] on the cytological and genetical changes induced by vegetable oils in *Triticum* established the mutagenic properties of the vegetable oils. Swaminathan also pioneered studies on the antagonistic and synergistic effects of different chemical mutagens assessed in terms of chlorophyll and viable mutation frequency in the M2 progenies of *T. dicoccum*. The *Emmer* wheat *T. dicoccum* was treated with different dosages of EMS and HA either individually or in combination. The authors [103] found that HA proved to be weak and EMS a potent mutagen both in normal plants as well as in an induced *chlorina* mutant. The frequency of mutations was very much reduced in combined treatments of EMS and HA. Swaminathan along with his students [104] demonstrated chromosome breakage induced in *Vicia faba* by monofunctional alkylating derivative of Acridine. Further, Swaminathan with his students Prasad and Krishnaswamy [105] showed that nitrosoguanidine is a potent mutagen. In the areas of chemical mutagenesis for mutation breeding of crop plants, one of Swaminathan's several interesting papers was published in *Mutation Research*. The title of the paper, *Enhancement of chemically-induced mutation frequency in barley through alteration in the duration of presoaking of seeds*, itself is suggestive of the involvement of DNA synthesis and occurrence of changes in the relative sensitivities of the 'G$_1$-S-G$_2$-M' phases of the prophase. This paper [106] has Dr. V.N. Savin of the Agro-physical Institute, Grazdanski St.14, Leningrad (formerly USSR), as one of the authors. Dr. Savin came from the erstwhile Soviet Union especially to work with Swaminathan who had already emerged as the one of the top world leaders in induced mutation breeding of crop plants.

Their findings [106] reported in *Mutation Research* are briefly presented below:

Treatment of barley seeds presoaked in water for different periods ranging from 8 to 40 h with EMS and N-nitrosomethyl urea (NMU) showed that the seeds were most sensitive to mutagenic treatments at 16 and 28 h from the beginning of pre-soaking. Autoradiographic studies revealed that the first incorporation of labelled thymidine occurs after 16 h of presoaking. The authors rightly relate it to the S-phase or the DNA synthetic phase which is normally of the longest duration in the cell cycle consisting of G1 (presynthetic gap) → S (DNA synthetic phase) → G2 (post-synthetic gap) → M (mitosis). Chemical mutagens are known to be more effective at the S-phase of the cells. The authors have rightly stated, *the lowering of sensitivity after 16 h and its subsequent rise after 28 h are more difficult to interpret, although the consistency with which the second peak occurs (at 28 h) makes it clear that there must be a basic cause for it.* The authors rule out a second cycle of mitosis being involved during the 28 h peak in sensitivity. The authors suggest four possible causes:

(i) Duplication of DNA and chromosomes at two different periods of a cell cycle, (ii) the existence of a sensitive phase during G$_2$, (iii) asynchronous but wave like DNA duplication both within a chromosome and between cells and (iv) appearance of some sensitising metabolic byproducts prior to germination. Of the possible four causes, a reference to a sensitive phase during G2 as early as 1968 reveals Swaminathan's intuition as well as his remarkable comprehension of not only the information already existing in a given field at a particular point of time, but also of those in emergence. Within the next two decades (i.e. 1988), the DNA repair and the role of *cell cycle check points* at the G$_2$ phase in the DNA repair and/ or apoptosis (i.e. programmed cell death) were more thoroughly elucidated. Considerable repair of the mutagens-damaged DNA occurs during the G2-phase. Swaminathan was absolutely right in implicating possible events specifically in the G2-phase for the observed spurt in sensitivity to the chemical mutagens at 28 h of presoaking of the barley seeds. Yet another interesting finding by Swaminathan [107] along with his doctoral research scholar George Varughese was that EMS is capable of inducing a wide range of mutations with a higher frequency than gamma-rays in the hexaploid dwarf wheat *Sonora-64*. Their data show gamma-rays are quite effective in inducing mutations involving the Q locus, but not in several other loci. The observation of interest is that the frequency of chlorophyll mutations with physical mutagen is negligible, while it is high with chemical mutagen. Swaminathan's mutation work in rice (*Oryza sativa*) reveals that the efficacy of chemical *vis-à-vis* physical mutagens varies with the crops as well. For instance, in their paper, *Mutation Breeding in Rice in India,* Swaminathan and his students [108] reported that chemical mutagens had no particular advantage over ionising radiation with reference to either mutation frequency or spectrum.

Special mention must be made of the brilliant contribution of Swaminathan and his student Siddique [109] to racial differentiation in *O. sativa* through mutational analysis. It is familiar to rice geneticists and breeders that based on a group of morphological differences and hybrid sterility, *O. sativa* has been sub-divided into three 'sub-species', viz., *indica, japonica* and *javanica*. Among these, the *indica* rice varieties are predominantly cultivated in India, Pakistan, Ceylon (Sri Lanka), Burma (Myanmar), Malaysia, Thailand, Laos, Cambodia, China and the Philippines, the *japonica* varieties in Japan, Korea, The Mediterranean region of Europe and California (USA), the *javanica* strains in Indonesia. There had been many studies on the genetic and chromosomal mechanisms involved in the origin and differentiation of *indica* and *japonica* rices and naturally, several divergent views exist. The relative roles of cryptic or visible structural changes in the chromosomes and sterility factors in the origin and subsequent differentiation of these two racial groups have not been precisely delineated. In their study, the authors [109] exposed the *indica* variety Taichung Native I (T.N.I) and the *japonica* variety *Taichung 65* (T. 65) to a wide range of physical and chemical mutagens. In the segregating (M2 and M3) population, several mutants each of which had the phenotypic characteristics of the other racial group were isolated. Interestingly, they found such mutants only in the M2 and M3 families derived from seeds treated with EMS and not in any of the radiation treatments. The breeding behaviour of the mutants was studied and stable mutants were isolated for the following key characteristics used in distinguishing *japonica* from *indica* varieties. The key characteristics pertained to colour of leaf, type of panicle, shape of grain, grain pubescence, resistance to alkali digestion, etc.

Led by Swaminathan, the conclusion of the authors [109] was as follows:

> *The study of mutants thus reveals that the constellation of characteristics which distinguishes japonica and indica rices may be affected either individually or in clusters. This suggests that the differentiation of these two racial groups does not involve a systemic or macromutation, but has probably proceeded through a series of independent mutations affecting grain and plant characteristics, brought together in a cluster probably under the influence of disruptive selection. If many of the mutant loci have an antimorphic effect in relation to the original allele, sterility could result in hybrid combinations. Disturbed genetic coherence rather than chromosomal differences, may be responsible for both the semi-sterility and skewed segregation ratios observed in the japonica-indica hybrid progenies. The possibility of changing japonica grains into indica type without altering the other characteristics affords an opportunity for making rapid use of good japonica varieties into the indica regions.*

By 1969, Swaminathan had reached the status of a living legend and global authority on mutation breeding and this was amply evident from his paper [110] published in *Proceedings of the XII International Congress of Genetics*. At the outset, he declared that plant breeding all over the world had entered a new phase with increased emphasis on the isolation of genotypes which respond to good agronomy and which have far superior and in some cases altogether new quality characteristics with respect to both nutrition as well as processing and cooking. He cited that Muller was of the view that the human ability to induce mutations in living organisms would liberate the humans from dependence on natural variability to produce desired crops. Swaminathan [110] stated: *Muller* [46] *in the very first paper on the artificial transmutation of the gene also expressed the hope that practical breeders need no longer be entirely at the mercy of existing variability, providentially supplemented on rare and isolated occasions by an unexpected mutational windfall.* It was true that Muller's optimism just did *not* come true at the level he expected. A review of the mutation breeding then of about 40 years reveals a few significant achievements, but the results by and large did not fulfil the general expectations owing to the rarity of progressive mutations. Also, polyploidy led to a masking effect on the phenotypic expression of induced mutations in several crop plants. Hence, some breeders came to the conclusion that mutation breeding may not be useful or even a supplementary method of plant breeding, let alone becoming a substitute for the Mendelian approach of hybridisation and selection of desirable recombinants. It is here Swaminathan [110] corrected the wrong notion by emphasising that *unlike the other methods of plant breeding such as exploitation of heterosis, where the practical exploitation phenomenon had not been handicapped by a lack of a theoretical understanding of the process itself, practical results in the field of mutation breeding will be proportional to the growth in our insight into the processes of induction, repair and recovery of mutations.* I strongly believe that this statement also emphasises Swaminathan's dictum to strengthen basic research in order to achieve an effective control over a technology for exploiting its true potential for better results. He himself covers a wide spectrum of disciplines in biology, biodiversity as also both basic and applied research.

He pointed out [111] to the fellow mutation breeders in the developed and developing countries that *the genetic variability existing in nature in crop plants and their wild relatives is largely conditioned by the nature and intensity of the selection sieves — natural and human — through which the population has passed during its phylogenetic history.* The process of evolution through 'blind' mutations acted upon by recombination and selection was aptly described by H.J. Muller as *muddling through.* Viewed in this background, Swaminathan drew attention to the new parameters of human selection. There are the obvious compulsions of population pressure, changes in the life-style demanding better quality processed foods and diminishing resources of soil, freshwater, biodiversity, renewable energy, etc. Of late, climate change has emerged as a major dominant challenge to sustainable agriculture as well as food and nutrition security. Table 1 on page 328 of Swaminathan's paper [110] presents a list of special characters of some varieties developed by mutation breeding. This table is reproduced as Table 3.4.

With his vast experience of the scale that few of his contemporary mutation breeders had possibly reached, Swaminathan in a scholarly manner not only summarised all the knowledge in the field that existed by 1969, but also showed the way forward. As usual, his approach is quite methodical and purely based on well-established scientific premise of facts. He effectively puts the facts and speculations into separate channels. So, his areas of coverage are as follows:

(a) Enhancement of induced mutation frequency

Since a majority of all mutations — both spontaneous and induced are of negative selection value, he emphasises that higher the mutation frequency the greater will be the chance of picking up economically useful mutations. He broadly suggested the following to achieve this:

- Varying not only the doses and dose-rates but also the radiations of different LET values.
- Equilibration of seed moisture content say to 2.5%, 3.5%, 6.8%, 10% during exposure to a series of gamma-ray doses. This suggestion reveals that Swaminathan was quite aware of the latest developments in the early 1960s that seed moisture, depending on its quantity in the seed, could act either as a radioprotector or a radiosensitiser. At very high levels of seed moisture (>10%), water facilitates recombination of the radiation-induced free radicals without much impact on the DNA. This would result in very few mutations. At about 3.0–4.0% seed moisture content, the free radicals have longer lifespan to react with oxygen (i.e. formation of ROS) and indirectly or even directly damage the DNA of the embryo of the seeds. This explains how the great mutation breeder of oats, A.T. Wallace [112], obtained a *mutational windfall* for resistance to the fungal disease (*Helminthosporium victoriae*), when the oat seeds of

Table 3.4. Special Characters of Some Recent Varieties Developed By Mutation Breeding.

Crop	Country	Name of Strain	Year of Release	Mutagen Used	Special Characters
Barley	Sweden	Pallas	1958	X-ray	Nitrogen response
Barley	Sweden	Mari	1960	X-ray	Earliness
Barley	U.K.	Midas	1967	Gamma-rays	Short Straw
Rice	Japan	Reimei	1966	Gamma-rays	Short-straw; greater yield stability
Triticum durum	Italy	Castelfusano	1967	Thermal Neutrons	Lodging resistance
T. aestivum	U.S.A.	Lewis	1965	Thermal Neutrons	Lodging resistance, productivity per day
T. aestivum	U.S.A.	Stadler	1965	-do-	-do-
T. aestivum	India	Sharbati Sonora	1967	Gamma-rays	High protein and lysine; seed colour
Gossypium barbadense	U.A.R.	Bahtim 110	1966	3:p	Free of Gossypol
Soybean	Japan	Raiden	1966	Gamma-rays	Earliness, short straw
Phaseolus-vulgaris	U.S.A.	Gratiot	1962	Gamma-rays	High protein; excellent canning quality
Phaseolus-vulgaris	U.S.S.R.	Saparke 75	1967	Gamma-rays	Suitability for mechanical harvesting

Source: Reproduced from Table 1 of Ref. [110, p. 328].

about 3.5% moisture content were exposed to gamma-rays of different dosages.

- Employing a wide range of chemical mutagens under diverse treatment conditions, the role of the DNA synthetic phase (S-phase) in responding specifically and varyingly to chemical mutagens, in particular, has already been discussed earlier.

(b) Recovery of induced mutations:

At the outset, Swaminathan [110] pointed out that *induction of mutations at the level of the locus and their subsequent recovery are two independent processes.* He elaborated that *the extent to which an induced mutation may survive and give rise to a mutant at the level of the organism is controlled by many factors among which the following relate to the invariate biological characteristics of the organism concerned.*

- Unicellular or multicellular.
- Sexual or asexual (vegetative) reproduction.
- Stage of differentiation of the primordial cells from which inflorescences arise, whether they are already present in the dormant embryo or arise later in the ontogenetic cycle.
- The number of primordial cells involved in the origin of each inflorescence.
- Genetic architecture of the organism, whether primarily diploid or polyploid of different types, which will determine the rigour of the somatic and gametic sieves.
- Characteristics of the loci involved in the mutation process.

So, the problems of recovery of induced mutations are manifold. A basic problem is that all crop plants are multicellular and hence the cells carrying mutation(s) will have to face competition with normal cells with regard to cell division, growth and survival.

Mode of reproduction whether sexual or vegetative (asexual) also greatly matters. Swaminathan points out that in sexually propagated plants, two types of selection would intervene before a mutation induced in a seed treated with a mutagen is expressed in the M2 generation. The first process occurs in the M1 somatic tissue — that is the *diplontic selection.* The second cycle of selection operates in the gametes of the M1 plants and is hence referred to as *haplontic selection.* Haplontic selection operates more rigorously in the pollen than in the ovules. Only a mutation that passes through both the somatic and gametic sieves will find phenotypic expression in the M2 and subsequent generations. Among the sexually propagated plants, the handling procedures are relatively simple in those which are hermaphrodite or monoecious than in dioecious plants. In a sexually or apomictically propagated plants, on the other hand, the somatic sieve is the only important one since the gametic sieve does not operate. Among apomicts again, there is a difference between those which are solely propagated through vegetative parts and those which are seed-propagated, without the meiosis in the cells, which gives rise to the functional megaspores.

Swaminathan emphasised that the vegetatively propagated plants present considerable problems in the recovery of mutation in view of the large number of meristematic cells present in the irradiated tissue and the consequent opportunities for cell competition and selection. Appropriate methods would need to be devised for providing opportunities for the cells carrying mutations which remain dormant owing to initial mitotic inhibition or delay to find phenotypic expression. The solution suggested by Swaminathan is chronic and recurrent irradiation. His design of the *Gamma garden* at the IARI was also to provide chronic and recurrent irradiation. Further, diplontic selection could be avoided in plants where seedlings can be raised by culturing single cells.

(c) Choice of mutagens

Swaminathan provided extremely valuable tips to the mutation breeders regarding the choice of mutagens between the physical and chemical agents. Based on the available data, he suggested that in polyploids, chemical mutagens like EMS which are capable of causing functional alterations in the gene have great advantage. Chemical mutagens were shown to induce chlorophyll mutations in autotetraploid barley, whereas the X-rays, gamma-rays and fast neutrons did not. From the extensive literature available on the subject, Swaminathan [111] summarised: *It is clear that there is as yet no definite indication that preference should be shown to either physical or chemical mutagens. Both have their value; neutrons among radiations and ethyl methane sulphonate among chemicals being generally the mutagens of choice. Professor Gustafsson's finding that some erectoid loci mutate more readily with neutron treatment and Dr. Jain's observation that HA and hydrazine give rise to non-overlapping classes of mutations in tomato and Drosophila emphasize the wisdom of using a wide range of mutagens in breeding experiments.*

Undoubtedly, some of the genetic loci respond to various mutagens in different ways. Also both physical and chemical mutagens often induce systematic mutations both in the tetraploid and the hexaploid crops. An interesting report by Upadhya and Swaminathan [113] was that EMS-induced *T. durum* ($2n = 4x = 28$) type of wheat in the M2 families follow-ing the treatment of the seeds of *T. pyramidale* ($2n = 4x = 28$). In the earlier studies of Bhaskaran and Swaminathan [114] involving ionising radiations in the tetraploid *Triticum,* viable mutations affecting a whole constellation of characters belonging to the category of macro- or sys-temic mutations had been extremely rare.

Yet, the fact remains that as emphasised by Swaminathan in several above-mentioned papers, basic science research and a very thorough understanding of the molecular and cellular response to various physical and chemical mutagens is absolutely essential to achieve significant degree of effectiveness and efficiency as well as 'directed mutagenesis'.

Hence, it is evident beyond any iota of doubt that Swaminathan's con-tributions to polyploidy and mutations towards achieving control over these evolutionary processes for achieving breakthrough in crop improvement have been truly phenomenal. His dominance and preeminence in these two areas of chromosome and genetic modification have come so naturally, largely because of his *holistic* as well as integrated approach to basic and applied research; his innate intellectual ability to launch inter-disciplinary research and development is rather unique. It is my view that several words and phrases such as *sustainable development, inter-disciplinary approach to solving complex biological problems,* and more recently *climate-resilient agriculture* are commonly spoken and even preached by several people, but only a tiny minority of them have really even tried to understand what these exactly mean, let alone their involvement with necessary diligence, devo-tion, intellect and competence to advance the frontiers of knowledge. Swaminathan stands glaringly apart from the 'crowd' of those who profess but not practice.

Chapter 4

Every Child Is a Scientist

The famous statement *every child is a scientist* implies that children are curious to know about everything around them. Often, they lose the sense of curiosity as they grow up and get engrossed with mundane things. There is no doubt that there are individuals who take to scientific research for the sake of livelihood, yet there are others who still maintain the curiosity of a child and therefore choose scientific research as a career. There is yet another category of scientists to whom science is the only thing that matters and they pursue science despite not getting a suitable job in the universities and research institutions. They even ignore the needs of their families. It was Carl Sagan the famous astrophysicist who once observed that curiosity is the driving force behind those who have made astonishingly great discoveries and inventions. Most children are intensely curious about everything — natural and human-centric innovations, around them. Galileo, Newton, Hooke and Raman are just a few to illustrate the fact that curiosity drove them to their discoveries. James Watt's invention of steam engine was also because of his curiosity. As children grow up, most of them get distracted from innate curiosity to mundane issues. Mental efforts are diverted to finding easy means to become rich and influential. Hence, as children grow up, their curiosity dies a silent death. Yet another category exists of professional scientists in national institutes or the universities who remain reasonably curious until they are into their 50s, and then their focus shifts to post-retirement plans. There is, however, a small proportion of the scientists and technologists with persistent curiosity that also progressively intensifies as they age as senior citizens. I continue to closely interact with a few such erstwhile nuclear physicists who are even now vigorously pursuing research and development in areas like 'cold nuclear fusion', radiation health effects at low and low-dose-rates, etc., despite their old-age problems, both social and financial. Some believe that their accumulated knowledge continues to need updating and also useful dissemination until they breathe their last.

Swaminathan is just like a child in the sense of his ever-increasing curiosity and enthusiasm. He analyses everything that he reads even today. To give an example, he commented just a few months ago that *gene editing* might be more acceptable than *genetic engineering*. I had a few serious questions which I thought I could take up with Swaminathan at his leisure. However, he himself told me a couple of days later that *gene-editing* has also unpredictable risks and uncertainties. This quality really sets him apart — that he could readily admit an error or an untenable postulate, and keep his mind open to corrections — and is the most admirable quality of a scientist. This is certainly not a trait of ambivalence as it is often mistaken to be. Biological systems are rather complex and have dynamic interactions with varying environmental factors. In such a myriad of complexities, the first postulate or even the first observation need not be so 'fixed' and 'immutable'. This situation is particularly true in the current appraisal of the health and environmental safety of the pesticide-producing genetically engineered crops. Today, more people realise the rDNA technology has several uncertainties and faulty assumptions than a decade ago.

The reason for the aforementioned prelude is to discuss his curiosity on the organisation of the 'chromosome' with histones, non-histone proteins, and DNA which is the main information-containing macromolecule. A decade after the elucidation of the structure of the DNA double-helix by Watson and Crick in 1953, cytologists wanted to understand the organisation of the DNA in chromosomes and chromatids (which are but one of the two strands of a chromosome). The chromatids and of course the chromosomes are visible under the light microscope which has a resolving power of about 2,500 Å or 250 nM. But the dimension of a DNA double-helix unit is just 20 Å. Therefore, the question remained on the structure, whether several units of DNA double helices were bundled together in parallel so as to reach the dimension of 2,500 Å. This view was, however, considered as inappropriate and possibly even irrational. This is so because several copies of the DNA double helices would greatly complicate the major functions of the DNA, viz., replication and transcription. Therefore, it is most likely that there is just only one copy of the DNA double helix per chromatid. This is the sort of conclusion that Swaminathan arrived at in 1964. His paper with his student Upadhya [115] is a commentary on the chromosome 'structure' revealed by simple techniques. In the early 1960s, there appeared two papers in *Current Science* [116, 117] that explained simple procedures such as fixation in 1N HCl (at 60°C) followed by suitable staining could help in revealing the fine structure of chromosomes at mitotic pro- and meta-phase. This is a view that was contradicted by several other cytologists, mainly Taylor [118]. At this juncture, Swaminathan and Upadhya [115] thoroughly examined the photomicrographs of the structure of the chromosomes revealed by the so-called *simple technique* and concluded them as products of artefacts, caused by the simple but excessively harsh treatment. They politely concluded: *studies of chromatic structure are difficult since the structural details are generally below the level of resolution of the light microscope and often above the size level at which, the thin sections required for resolution in electron microscopy can give a complete picture.* A few years later, Swaminathan and his student Bastia [119] followed an ingenious method and produced very clear electron micrographs of the ultrastructure of the interphase chromosomes. In their introduction, they point out that *electron microscopy of this sectioned material had been of very limited value in unraveling the organization pattern of chromosomes.* They discuss the difficulty to interpret the complex three-dimensional structure of the chromosomes out of the essentially two-dimensional pictures. Swaminathan and his student [119] largely followed the modification in the technique made by Professor Gall [120] and investigated the ultrastructural organisation of the interphase chromosomes of frog erythrocytes and human leucocytes. Their paper gave a scientifically precise suggestion that the 500 Å microfibrils were nucleohistone in nature. Their findings of the uncoiling of microfibrils brought about by treatments with 1 M ammonium acetate and trypsin supported the role of histones in the supercoiling of DNA. Their work in 1967 along with those of DuPraw [121] has led to the firm demonstration that each chromatid has only one DNA double helix unit (i.e. 'uninemic') and the DNA with different types of histones is organised into 'nucleosomes'.

From the aforementioned account of Swaminathan's scientific contributions, there could be no disagreement, whatsoever, that from being as inquisitive as a child, he quickly shot up as the global leader in several areas of Life Sciences. The term *Life Sciences* had not been coined when he was studying the phenomena and processes of living beings in an integrated and inter-disciplinary manner that was intrinsically the Life Sciences. He had many things in his mind; he wanted to know the ultra-structure of the eukaryotic nuclei and chromosomes; he wanted to elucidate the mitotic division process in a lower eukaryote, the yeast cells; he wanted to elucidate the stage during which DNA synthesis began and ended; he then jumped across several steps to unravel whether polyploidisation of nodule cells in the leguminous plants is the 'cause' or 'effect' of infection by rhizobia; then he switched back to determine the chromosome(s) and the genome responsible for increased radiosensitivity of the hexaploid wheat (*T. aestivum*) — all these were after the major contributions to cytogenetics of cultivated potato and its wild diploid and polyploidy relatives. Not to forget, he was also a radiation biologist *par excellence*; he was among the first very few to relate radiation dose with radiation quality (i.e. LET, RBE, etc.) and the *oxygen enhancement ratio* when cells and organisms are exposed to low LET ionising radiation such as gamma-rays and 250 kVp X-rays. He then made a breakthrough in understanding the mechanisms of radiobiological pathways by segregating the radiobiological pathways into 'direct' and 'indirect' effects. In the area of radiation genetics, especially for its application to mutation breeding, Swaminathan is amongst the very few to have initiated appropriate studies to understand and define the physical, chemical and environmental factors which influence the radiosensitivity in a varied manner. He emerged as a stalwart leader in emphasising and demonstrating that success in mutation breeding required a good theoretical knowledge derived from basic research, whereas success in a heterosis breeding could easily come without such basic knowledge in integrating biology, physics and chemistry. He also goes further to point out the tenacity of the life processes, particularly the role of the DNA in initiating various modes of repair when it was damaged by mutagens; he then made elegant and extremely meaningful classifications of the induced mutations as macro, micromutations and systematic/systemic mutations. His mutation work in *indica* and *japonica* rices is extremely exciting since he demonstrated the occurrence of mutants of *japonica* with *indica* traits and *vice versa*.

There is no rhyme or reason for anyone to have ever been petty and jealous over his accomplishments because he remained a child in the sense of closely guarding his sense of curiosity. It is probably true that *every child is a scientist*, but Swaminathan has shown that a child in you does not have to die, but could go on growing in its curiosity. So, this intellectual giant, one who has used sciences for social and gender welfare, is a *91-year-old child* for his insatiable curiosity and thirst for knowledge. He is remarkably amenable to 'paradigm shifts' necessitated by new knowledge and deeper insight. He is not at all dogmatic as many who have risen to high offices and influential positions are. More recently, he has analysed and come to the conclusion that genetic engineering with all its known ills and uncertainties should not be continued with; instead, more benign technologies such as the 'Marker-Assisted-Selection' should be pursued. As it would be discussed in later chapters, he is absolutely firm in his view that technologies which are antithetic to ecological and social sustainability have no place in the planet Earth that is now at the crossroads.

There are many who wonder if he was elected to the Royal Society, London (FRS) for ushering in the Green Revolution to India. The major purpose of this biography is to bring out and put together briefly the high quality and intensity of his innovative basic research in cytogenetics, cytology and radiation and chemical mutagenesis and to establish that these indeed formed the basis for his election to the 'Fellowship of the Royal Society' (FRS) in 1973 when he was just 48 years.

He was known and recognised as a born Life Scientist who devoted the later years to food and nutrition security. No doubt, he specialised in genetics and cytogenetics, but he was also very good in plant physiology, plant pathology and mycology, entomology, soil science, agro-ecology, and economics, particularly ecological economics (which is distinct from environmental economics). It is his command over a very broad spectrum of natural and social sciences that has made him a true 'Father of Do Ecology'.

Swaminathan's Role in the Genesis of the School of Life Sciences (SLS) of the JNU:

The first Prime Minister of independent India, Pandit Jawaharlal Nehru died in 1964 and within the next couple of years, the Parliament of India enacted a bill to set up a Central University in New Delhi as a living tribute to him. The decision was that University should not be just 'one more University' in the country but one that would fulfil Nehru's vision and dream of an ideal University — one that would indulge in 'adventure of ideas' and be 'socially relevant'. This new University was christened the *Jawaharlal Nehru University (JNU)*.

In the creation of the academic programmes of the JNU, the primary step was to appoint a Vice-Chancellor with enormous wisdom, leadership and one who spontaneously commanded the respect of one and all. One who qualified in all respects of ideal virtues and had an inspiring personality was late G. Parthasarathi (affectionately called 'GP'). He presided over the Academic Advisory Committee of the JNU and constituted the following Working Group to initiate the creation of the School of Life Sciences with broad focal themes of research and teaching programmes at the M.Phil. and Ph.D. levels and also identify the core faculty members. The Working Group consisted of the following:

1. Professor M.S. Swaminathan — Chair
2. Professor M.G.K. Menon
3. Professor T.S. Sadasivan
4. Dr. A.R. Gopal-Ayengar
5. Professor V. Ramalingaswami
6. Professor O. Siddiqui
7. Dr. G.P. Talwar

The Working Group met on December 10, 1969 and discussed the inter-disciplinary programmes of academic excellence and national relevance. The Working Group also met with Professor George Beadle, Noble Laureate who was in New Delhi on that day and had made two general yet essential suggestions:

(i) To avoid rigidity of conventional departments
(ii) To lay greater emphasis in faculty appointments on the qualified individuals who are good in group interactions

Taking the cue from Professor Beadle's statement, Swaminathan emphasised that individuals with *genes for co-operation and team work* should have preference in appointment to the faculty positions and that *isolated islands of research activities* should be discouraged. In support of these views, it was reasoned that the contemporary modern biological research had moved away from the descriptive phase and had entered the realm of elucidating the processes and functions which integrate physics, chemistry, structural biology, biophysics and biochemistry with all these emanating in one way or another from molecular signalling, gene expression and epigenetic control. Whether the issue involved is the photosynthesis, or vision or immune response, there are the inevitable roles of physics, physical chemistry and structural biology related to physiological functions controlled by genes. The Committee identified immunology, photobiology, radiation biology, neurobiology as some of the areas to initiate the inter-disciplinary research and teaching at the post-graduate levels.

'GP' has done the right thing in requesting Swaminathan to be the Head of the Working Group, the members of which are undoubtedly stalwarts in their own but rather 'narrow' areas of specialisation; they were not so well committed to or experienced in inter-disciplinary research as Swaminathan. To the readers of this biography, the earlier two chapters should be quite reassuring that Swaminathan, unlike most others of his days was driven by curiosity and that led him to get interested in problems from several areas in biology. This also established that he was not a traditional agricultural scientist confined to one of the disciplines or sub-disciplines, but holistic in his approach.

The strong foundation he laid for the academic programmes in the School of Life Sciences, JNU is indeed the major cause of its achieving international recognition. In February 2014, a grateful School of Life Sciences felicitated Professor Swaminathan and other living members of the Working Group for having established not just a foundation but a 'culture' of inter-disciplinary approach to solving the unsolved problems in biology, medicine and agriculture. It is this 'culture' that is passed on unfailingly from the departing old to the incoming new faculty. There is the natural turnover of the faculty, but the 'culture' of inter-disciplinary research and teaching remains strong and irreplaceable. It will remain in the annals of the SLS and in all other centres which practice inter-disciplinary Life Sciences that the Father of India's Green Revolution, the prime architect of the Evergreen Revolution is also the founder of inter-disciplinary research and teaching programmes in the Life Sciences.

Swaminathan is more known for ushering in food and nutrition security of the nation through the Green Revolution, but that was five decades ago. His journey in science started in 1950. Hence, the preceding chapters brought out his scientific contributions from 1950 until mid-1960; during 1967–1968, Swaminathan with immense curiosity about nature and her ways, had become laden with the tremendous responsibilities to find ways to generate adequate food for millions of new mouths added annually, while the ecological foundations of sustainable agriculture was eroding progressively.

Chapter 5

Child-like Curiosity Matched with Social Responsibility

Sir C.V. Raman too had insatiable curiosity to know everything on this planet as Swaminathan. Alike in their curiosity, they were also different in their outlook, particularly regarding the social purpose of science. To Raman, nothing else but his field of science mattered; his intense involvement in a specialised area of scientific pursuit had confined him to an 'ivory tower'. Swaminathan, despite having had rather close association with Raman, chose to be a people's scientist. He extended his hand in warmth to interact with social scientists. The expected natural obligation on the part of a people's scientist is to establish a *social contract of science*. Considering the dismal food security scenario at the time of the independence of India in August 1947, and thereafter for the next couple of decades, the most appropriate social contract of science was to increase agricultural food production and *availability* of food mainly through the enhancement of the productivity of the crops as also the *access* to food by all people. With Indian agriculture having been a 'gamble with monsoon', droughts and famines have been a regular feature; the great Bengal famine of 1943–1944 loomed large as backdrop to India's independence in 1947. To India's first Prime Minister, Pandit Jawaharlal Nehru, it was a great concern that should a famine of the magnitude of the Bengal famine recur, the faith in self-determination of the people of young Independent India would be devastated. Pandit Nehru also understood that if agriculture in India would fail, nothing else in the domains of science, technology, education, industry, commerce, foreign policy and so on would ever succeed. Having been deeply conscious of this bottom line, he put agriculture and self-sufficiency in food at the top of the national agenda and proclaimed, *everything else can wait, but not agriculture.* Following this, several measures to stimulate food production, including land reform, irrigation, fertiliser production, strengthening of research and organisation of a national extension service were initiated in the 1950s. More land was brought under farming and augmentation of irrigation sources was also done. There was an increased production of wheat and rice, but productivity per unit area of land remained quite stagnant. The research goal was to enhance the *per plant productivity* with high levels of plant nutrition. Hence, late Dr. K. Ramiah in 1952 started a programme for incorporating genes for fertiliser response from *japonica* rice varieties to *indica* rice varieties at the Central Rice Research Institute (CRRI), Cuttack under sponsorship of Food and Agriculture Organization (FAO) and Indian Council of Agricultural Research (ICAR). Swaminathan worked in 1954 in this project. The main aim was to select from segregating populations of *indica* X *japonica* crosses, lines which showed the ability to utilise nitrogen effectively about 100 kg/ha. This approach resulted in a yield of about 5 tonnes/ha. There were a few unexpected genetic problems rendering the speedy selection of high-yielding rice varieties from *indica* X *japonica* crosses difficult. Yet, a few varieties like ADT-27 and *Mashuri* cultivated in Tamil Nadu and Malaysia, respectively were developed.

As stated earlier, Swaminathan did not continue with CRRI, Cuttack and he joined the Botany Division of the Indian Agricultural Research Institute (IARI), New Delhi in 1954, after a few months of his joining the CRRI. He brought with him from the CRRI the idea of developing non-lodging and high level fertiliser responsive varieties of wheat on the lines of the work earlier initiated in rice. All the locally adapted best varieties of wheat were growing tall and easily lodged when heavy doses of nitrogen were applied to the soil. Swaminathan initiated three different plans of action to develop short and stiff-straw plant types, with normal long panicles, so that each plant would have a large number of well-filled grains. These three different strategies were as follows:

I. Hybridisation and selection involving cultivated bread wheat and the semi-dwarf stiff-straw *compactum* and *sphaerococcum*, sub-species of *T. aestivum*.

II. Attempts were made to induce *erectoides* mutants in the commercial varieties through the use of radiations and chemical mutagens. Although discussed already in great detail, it is again emphasised that Swaminathan and his team carried out extensive basic research on the mechanisms of action by the physical and chemical mutagens as well as the factors influencing the frequency and spectrum of induced mutations with the objective of realising desired types of mutants. Unfortunately, the success in this area was highly limited.

III. To increase the straw stiffness through different chemical treatments. The reason why straw stiffness became such an essential prerequisite for favourable response to water and fertiliser was the tendency, among the cultivated tall wheat varieties, to lodge or fall down when optimal level of fertiliser was applied. Also, such lodging made it quite diffi-cult to irrigate during grain development phase when the crop will benefit much from water availability. Swaminathan pointed out that with the earlier tall varieties, it was difficult to get economic response to the application of mineral fertilisers and adequate irrigation. He also explained that breeding of non-lodging varieties was accorded such high priority during the 1950s, when the country had taken to the path of expanding area under irrigation and manufacturing mineral fertilisers.

Coming back to the three strategies, it was unfortunate that all the three of them failed because short and stiff-straw was always associated with short panicles and consequently, each plant had fewer grains and the yields stagnated at less than 1 tonne/ha. The genetic basis of inseparable association between short and stiff-straw character and the short panicle length was ascribed either to close (tight) linkage between these two traits, or the phenomenon of *pleiotropism*, in which the function of one gene exerts influence over the function of unrelated genes.

In his Presidential Address to the Agricultural Sciences Section of the *55th Indian Science Congress Meeting* held in January 1968 at Varanasi, Swaminathan [122] said: *The discovery in Japan of genes in the Norin wheat variety, which confer a dwarf and non-lodging plant habit, hence opened the door to the reconstruction of the morphology of the wheat plant. Several dwarfing genes were known for a long time in wheat, such as the S or C loci which govern the sphaerococcum and compactum characteristics, respectively. These loci, however, have a pleiotropic effect on the ear, making it very dense and compact. The first variety which appeared to have the desired combination of short plant height, lodging resistance and kernel type was 'Norin-10'. This variety was one of a collection of Japanese wheats brought to the United States by Dr. S.C. Solmon in 1948. Three recessive genes for dwarfing, with additive effect, have so far been identified.*

Swaminathan implied that though his three different approaches to get a wheat plant with desired morphology did not succeed, the 'nature', gifted, meaning spontaneous genetic changes and initially 'selection by nature' and then a selection possibly by the breeder Dr. Gonziro Inazuka in Japan did what humans could not.

Swaminathan has been among the fewest of the few cytogeneticists and plant breeders who have been vigorous crusaders of conservation of biodiversity. It is loud thinking on my part that Swaminathan's innate passion for conservation of nature should have become substantially accentuated by his realisation that nature provided the *Norin-10* with the desired morphology, which he could not produce in laboratory despite having tried three different methods. Swaminathan's deep involvement and unparalleled service to the conservation of natural resources in general and biodiversity in particular is dealt with in detail in a later chapter. Swaminathan's [123] magnificent Presidential Address at the XV International Congress of Genetics, held in New Delhi in 1983, brings out the importance of biodiversity conservation for future food and nutrition security. Of course, several developments have been taking place towards conservation of biodiversity following of his most powerful address. These aspects are discussed in Chapter 10.

Chapter 6

Agricultural Transformation

6.1 Preamble

There was never ever a doubt that India's food security was extremely precarious at the time of India's independence and couple of subsequent decades. The situation of food insecurity since India's independence was gradually worsening because of at least three reasons: (i) India's population was increasing steeply, (ii) the ecological foundations of agriculture was progressively undergoing degradation, and (iii) monsoon was becoming increasingly erratic.

Around the year 1950, India's human population was about 359 million (~36 crores) and India's food grain (rice, wheat, coarse grains and pulses) production during 1950–1951 was 50.8 million tonnes. During the next 10 years (1959–1960), India's food grain production went up to about 82 million tonnes, (an increase of 62%) whereas the population soared to about 450 million (an increase of about 80%). Clearly, the rate of population growth was far ahead of the rate of growth of grain production. This trend, highly disturbing, would have gone on but for Swaminathan's introduction rather deliberately of what he called the 'intensive agriculture'.

Swaminathan was certainly aware of the negative aspects of the intensive agriculture, yet he had to embark on it. He knew that failure of monsoon consecutively for a couple of years would lead to near famine situation. In any case, India was importing about 11 million tonnes of wheat and rice under PL 480 from the US. Apparently the grim situation in the food front triggered Ehrlich [124] to write the book, *The*

Population Bomb (1968). In this original edition of 1968 (his 1990 edition was more moderate), the beginning statement was: *The battle to feed all of humanity is over. In the 1970s hundreds of millions of people will starve to death in spite of any crash programs embarked upon now....* The reference in the book was directly to India. This book predicting hunger and death of hundreds of millions of people was soon followed by another book by Paddock brothers [125], *Famine 1975! America's Decision: Who will survive?*, reportedly a best-seller. The authors William and Paul Paddock (the Paddock brothers) described the rapidly growing population of the world, and a situation in which they believed it would be impossible to feed the entire global population within the short-term future. They asserted that widespread famine would be the inevitable result by 1975. The book noted the following points:

1. The underdeveloped nations have exploding populations and static agricultures.
2. The 'Time of Famines' will be seriously in evidence by 1975, when food crises will have been reached in several of these nations.
3. The 'stricken people will not be able to pay for all their needed food imports. Therefore, the hunger in these regions can be alleviated only through the charity of other nations'.
4. The only important food in famine relief will be wheat, and only the United States, Canada, Australia, and Argentina grow significant amounts of wheat.
5. The United States, the only one of these four countries that has historically given wheat to hungry nations, is the 'sole hope of the hungry nations' in the future.
6. *Yet, the United States, even if it fully cultivates all its land, even if it opens every spigot of charity, will not have enough wheat and other foodstuffs to keep alive all the starving.*
7. *Therefore, the United States must decide to which countries it will send food, to which countries it will not.*

Under these conditions, they suggested *triage* approach to solve the problems. The word *triage* is a medical term, and it is not found in the Oxford dictionary. *Triage* is used for sorting of and allocation of treatment to patients, especially battle and disaster victims, according to a system of priorities designed to maximise the number of survivors. Paddock brothers used the *triage* to classify underdeveloped countries into three categories: (1) Those so hopelessly headed for or in the grip of famine (whether because of overpopulation, agricultural insufficiency or political ineptness) that our aid will be a waste; these 'can't-be-saved nations' will be ignored and left to their fate, (2) Those who are suffering but who will stagger through without our aid, 'the walking wounded', and (3) 'Those who can be saved by our help'.

The Paddock brothers [125] were aware that their policy of abandoning food aid to the 'hopeless countries' (India and Egypt for example) would lead to an immediate worsening of the situation there, but they wrote *to send food is to throw sand in the ocean*. Using the *triage* system they hoped to avoid a broader catastrophe and stabilise the global population.

These two books added insult and humiliation to the hungry, food insecure nation. The compulsion for Swaminathan was to limit his immense curiosity for basic research and divert his energy, time and resources to meet the basic needs of his people, particularly food and access to it. About the same time (late 1950s to early 1960s) his three approaches to develop 'short, stiff-strawed, normal panicle length' wheat plant variety was not fruitful, he read about *Norin dwarfing genes* and about Dr. O.A. Vogel in the Washington State University using these genes and breeding a new dwarf winter wheat variety, *Gaines* that produced a several-fold increase in the yield. The *Gaines* variety favourably responded to heavy dressings of fertiliser and copious irrigation 'without' lodging — these plants stood erect at the time of harvesting. It is reported that the first semi-dwarf *Gaines* variety produced record yields, about 11, 196 kg/ha in 1965.

Impressed with the performance of *Gaines*, Swaminathan wrote to Dr. Orville Vogel for some seeds of the 'miracle' wheat. Swaminathan [126] writes, *Dr. Vogel was kind enough to send me seeds of Gaines, a semi-dwarf winter wheat variety with red grains. He further suggested that I should approach Dr. N.E. Borlaug in Mexico for seeds of semi-dwarf varieties in a spring wheat background (winter wheat needs long hours of sunlight to flower and set seeds)*. Under the north Indian conditions during winter, the daylight is short and therefore the *Gaines* variety with winter wheat background would not do well.

That was the turning point for a long Swaminathan–Borlaug association and genesis of an 'intensive agriculture' that was given a name, the Green Revolution.

6.2 Role of 'Professor M.S. Swaminathan Library' in Providing Knowledge Power to Identify the Seeds and Its Sources for Green Revolution

In the earlier sections, three aspects were highlighted:

 (i) High level fertiliser-responsive wheat varieties were required.
 (ii) Three different scientific methods to produce wheat plants having short culm with stiff straw but normal length panicle did not result in tangible success.
(iii) Food-insecure situation coupled with humiliating description of India's food front and chances of human survival (reference to books [124, 125]) had touched the sensitivity of the people of India. Humiliation was far worse than hunger. Yet, the unfortunate situation was that laboratory research did not result in an ideal wheat plant type that was dwarf or semi-dwarf, that had stiff-straw and long (normal length) panicles.

It was at this juncture that Swaminathan's regular habit of making use of the journals, books, bulletins, monographs and Newsletters came in good stead. The Indian Agricultural Research Institute (IARI) library that was dedicated in the name of Professor M.S. Swaminathan on the April 29, 2016 has an extremely valuable collection of scientific journals, publishing original and review articles on almost every facet of agricultural sciences in general, cytogenetics, genetics and plant breeding in particular (Fig. 6.1). I have gathered from Swaminathan's article on Green Revolution that he had read and cited the work of Dr. Vogel and his associates with semi-dwarf hexaploid wheats. This paper entitled, *Semi-dwarf growth habit in winter wheat improvement for the pacific northwest* [127] is cited by Swaminathan and his co-authors [128] in their paper in *Zeitschrift fur*

Figure 6.1. IARI Library dedicated to Professor M.S. Swaminathan in April 2016.

Pflanzenzuchtung. Furthermore, Dr. Vogel was his good friend and both have had long discussions on the dwarf and semi-dwarf wheat varieties and related matters.

So, it was indeed Swaminathan's reading habit that led him to trace Vogel's 'wonder wheat'. Then it was his close association and friendship with Dr. Vogel that facilitated in getting some seeds of *Gaines* wheat, and especially Vogel's advice to Swaminathan to approach Dr. Norman Borlaug to obtain some seeds of dwarf and semi-dwarf spring wheat varieties which would be more suited to North Indian plains. The main concern was the day length consideration. Vogel's dwarf and semi-dwarf winter wheat varieties required long day length, but in the north Indian plains where wheat is grown in winter ('rabi' crop), the day length is short and inadequate. Consequently, the flowering and seed setting would be impaired.

Vogel's suggestions were well taken by Swaminathan, who immediately contacted Norman Borlaug. Borlaug and Swaminathan had known each other. Borlaug immediately responded to Swaminathan's request for

Mexican dwarf and semi-dwarf wheat seeds. The Mexican dwarfs and semi-dwarfs did outstandingly well in the North Indian wheat belt. The wheat yield revolution in India became the most significant accomplishment in the post-independence era. In 1968, I was in Canada when the Government of India commemorated the great achievement with the release of a Special Stamp showing the wheat yield revolution and the IARI Library (now Professor M.S. Swaminathan Library) in the backdrop. I was so proud to have been the student of Swaminathan, the architect of wheat yield revolution first, then of rice and other crops.

Although the Nobel Peace Prize should have been shared jointly by Borlaug and Swaminathan, only Borlaug alone was proposed for the Green Revolution in India. Borlaug's several subsequent interactions with Swaminathan give me an impression that he (Borlaug) must have deeply felt that equal justice was not done, and Swaminathan was rather unfairly excluded from the nomination. After receiving the Nobel Peace Prize, Norman Borlaug acknowledged Swaminathan's initiative and prominent role in the ushering in of India's Green Revolution in a letter to Swaminathan. The relevant section of Borlaug's letter addressed to Swaminathan reads as follows: *to you, Dr. Swaminathan, a great deal of the credit must go for first recognizing the potential value of the Mexican Wheat Dwarfs. Had this not occurred, it is quite possible that there would not have been a Green Revolution in Asia.* Borlaug was absolutely right about what he has referred to Swaminathan's contribution. In the magnitude of various factors which culminated in the realisation of the Green Revolution, Swaminathan's role is just second to that of 'nature' that gifted to humankind the 'dwarfing genes' for wheat in Japan (i.e. *Norin dwarfing genes*) and for rice in China (*Dee-Gee-Woo-Gen*). There is also an irony in the sense that Norman Borlaug was a plant pathologist (who studied the different 'rust' diseases in wheat) whereas Gunziro Inazuka, Orville Vogel and Swaminathan are geneticists more directly connected with plant breeding. This indeed was not the only issue; even more serious was the long-term ecological, social, gender and economic impacts of the Green Revolution technology itself. The new semi-dwarf and dwarf wheat varieties which responded well to applications of heavy doses of chemical fertilisers and copious irrigation were clearly known to be potentially capable of causing degradation of soil, freshwater and biodiversity and therefore, in the long run, the yields would taper off. Many critics of the 'new agriculture' had put Borlaug, the Nobel

Laureate and Swaminathan at par in jeopardising the future agricultural prospects of India. Many of them including Dr. Vandana Shiva attacked Swaminathan; it later turned out that they were all barking at a wrong tree. In this regard, reference is made to [122] where Swaminathan had cautioned against *exploitative agriculture*.

Swaminathan was indeed the very first scientist, in fact, regarded as the *Father of Green Revolution* to describe the new high-yielding agriculture as *exploitative agriculture* and it would not be suitable for sustainable agriculture. Even now many people might not know that he had described it as an *exploitative agriculture*, and therefore, it should be used only to get a breathing space, and that it should be soon replaced with a more sustainable agriculture — he said these about seven months before the jubilant nation and the Government of India released a Special Wheat Revolution Commemoration Stamp on the July 17, 1968.

Addressing [122] the Agricultural section as its President at the *55th Indian Science Congress Session* in January 1968 at Varanasi, Swaminathan made several thoughtful remarks which would be valid and relevant for a long time to come. These remarks and analyses reveal Swaminathan as a scientist who from the very beginning integrated the elements of ecology, genetics, soil science, plant pathology, social and gender dimensions of sustainable agriculture and above all as an honest critic and appraiser of the *new agriculture* of which he was the chief scientist, and architect. Norman Borlaug did not initiate, but he was a willing accompanist in India's *new agriculture* based on wheat plants tailored to utilise the mineral fertilisers and copious irrigation applied to the field, and also the solar energy in photosynthesis, to produce more grains per earhead/panicle. It is gathered that India got the *Norin-10* dwarfing genes only after Swaminathan requested for them first from Orville Vogel, and then from Norman Borlaug in 1962, whereas Pakistan had already been delivered these 'miracle wheat seeds' a couple of years earlier by Borlaug himself. One can only wonder as to the reasons for the *new agriculture* to not take root, in Pakistan.

The first point that emerges prominently is that Swaminathan's appraisal and critical evaluation of the *new agriculture* might not have been appreciated by Borlaug; in fact, he might have greatly resented it. There is no direct evidence, however. Yet, there are many ways of judging Borlaug's attitude. Some people believe that he was anti-ecology and anti-environment. There is evidence that he had made rather uncharitable remarks and untenable criticisms of Carson [129] who wrote the book *The Silent Spring*. As those who have read this book would know, Carson had shown the devastating effects of DDT and other chemical insecticides on the non-target organisms such as pollinators, butterflies, beautiful birds, etc. In support of the DDT and the chemical pesticides, Borlaug attacked her for her 'environmentalism' that impedes the goal of achieving higher productivity and total production. Superficially, Borlaug might appear to have been justified with his single-track mindedness to enhance the global availability of cereal grains, but an in-depth analysis unambiguously reveals Borlaug's lack of appreciation of the role of ecological foundations for sustainable agriculture. In fact, in his lifetime, Borlaug could not have missed knowing about the fatigue of yield gains associated with Green Revolution in India. He would also have known that Green Revolution which ensured food security at the national level failed to establish the same at the individual household level of hundreds of millions of rural poor. The scenario depicted a paradox of 'mountains of grains on one hand, and millions of hungry people on the other'. In fact, the adverse effects of chemical pesticides on the pollinating insects including honeybees, reflected on poor fruit set of vegetable and fruit crops which depend on insect pollinators. The fact of a mutual inter-dependence and symbiotic relationship among living organisms in planet Earth's biosphere had been known since the days of early civilisation.

The foregoing forms the basis of the dichotomous views on the *new agriculture* (i.e. Green Revolution) between Borlaug and Swaminathan. With this prelude, relevant statements/remarks made by Swaminathan [122] in his Presidential Address to the Agricultural Section of the *55th Indian Science Congress Session* in January 1968 at Varanasi are as follows:

(i) *We are now on the threshold of a major agricultural transformation, shifting from what I would like to term "natural" into "exploitative" agriculture. In natural or static agriculture, the cultivated plants largely depend for their survival and productivity on natural factors supported by minimal assistance from man. Exploitative agriculture, on the other hand, is based on an efficient use of both natural as well as artificially created resources from bringing about a continuous rise in the yield and income obtained from unit area without detriment to the long-term productivity of the land. Developments in genetic engineering including the technique of algeny or artificial transmutation of genes, in chemical technology including the synthesis of fertilizers, hormones, pesticides, fungicides, antibiotics and herbicides and in engineering including water technology and the design and production of more efficient farm machinery, are all responsible for ushering in the era of exploitative agriculture.*

(ii) *In the wild state, the plant has to fend for itself and hence possesses certain in-built mechanisms for propagation and protection. These mechanisms, which are necessary for the survival of the plant in nature, become superfluous under domestication and are also not conducive to maximum productivity. The attributes desirable for productive agriculture are in many cases just the opposite of that needed for survival under nature. For example, synchronous tillering is a very valuable character in cereals and millets, while a highly asynchronous tillering is of positive selection value under natural selection.*

(iii) *The pattern of improvement in crop production in most countries reveals a discontinuous rise, associated with some major advance in crop husbandry. For example, the yield of wheat in the United Kingdom during the period 1400–1750 remained at the level of about 7.5 bushels per acre. In about 1750, the principle of crop rotation was introduced into agriculture and this resulted in the per-acre yield increasing from about 7.5 to 20 bushels. The highest rice yield in the world occurs in the Murrumbidge region of Australia, where rice is grown only once in five years, the land being under legumes during the other 4 years.*

In this Presidential Address, he also presented an analysis of the causes of malady of poor yield in rice, and the necessary remedial measures:

As of 1960s, the average rice yield in India was about 1.1 tonnes/ha, in contrast to 4 tonnes/ha in Japan. The Indian rice varieties are subspecies of *indica* of *Oryza sativa*, Swaminathan identified some of the reasons for the low yield in rice as follows:

(a) The weak and tall straw of the *indica* varieties makes the cultivation of rice under good conditions of soil fertility and the application of fertilisers difficult without causing lodging, (b) poor photosynthesis due to extensive cultivation of rice during the monsoon when the sky is cloudy during most parts of the day, (c) poor utilisation of sunlight due to the shading of the lower leaves by the upper ones, (d) poor soil and water management, and (e) lack of attention to details in cultural practices such as spacing and depth of transplanting.

In the context of overcoming some of the problems mentioned above, Swaminathan referred to a spontaneous mutant in the variety *Dee-Gee-Woo-Gen* with the characteristics given below:

(a) a dwarf plant habit, the plant attaining a height of about 60 cm, (b) stiff and erect leaves enabling maximum interception of sunlight, (c) insensitivity to photoperiod enabling the cultivation of crop in any possible season and (d) absence of seed dormancy, rendering sowing possible immediately after harvest. Using this mutant, scientists in several parts of the tropics developed fertiliser-responsive and photo-insensitive varieties, which revealed enormous possibilities for increasing the yields of *indica* rices.

Referring to the performance in the yield increases of the dwarf and semi-dwarf Mexican wheat derivatives in India, Swaminathan [122] presented the table containing data on productivity per day in some wheat varieties (Table 1, p. 239 of the Proceedings of the *55th Indian Science Congress*, Part II, Presidential Address) (Table 6.1).

Sonora-64 is an early variety and is well-suited for being grown in rotation like maize–wheat, potato–wheat, rice–wheat, sugarcane–wheat, etc. Table in Ref. [122] provides data on wheat yield in rotation, involving, late sowing (Table 6.2).

Table 6.1.   Productivity per Day in Some Wheat Varieties.

Variety	Days in the Field	Productivity (kg/ha/day)
NP880 — Local tall variety	151	17.1
Lerma Rojo — late variety	153	24.8
Sonora 64 — early variety	133	42.5

Source: Data from Table 1 of Ref. [122, p. 239].

Table 6.2. Wheat Yield in Rotations Involving Late Sowing.

Location	Rotation	Wheat Variety	Date of Sowing	Yield Q/ha
Delhi	Jowar–wheat	*Sonara 64*	29-12-64	42.4
Jullunder	Potato–wheat	*Sonara 64*	02-01-66	40.9
Jullunder	Potato–wheat	*Lerma Rojo*	02-01-66	36.2
Samastipur	Rice–wheat	*Sonara 64*	05-12-65	37.1
Darbhanga	Rice–wheat	*Sonara 64*	19-12-65	25.5
Delhi	Rice–wheat	*Sonara 64*	04-12-65	55.3

Source: Data from Table 1 of Ref. [122].

Having covered a wide range of issues both positive and negatives as well as what are feasible, and what are not, Swaminathan brilliantly narrated the risk of adopting the *exploitative agriculture* on long-term routine basis. What he [122] had stated towards the end of his address is reproduced below:

I mentioned at the outset that we are now entering an age of exploitive agriculture. Exploitive agriculture offers great possibilities if carried out in a scientific way but poses great dangers, if carried out with only an immediate profit or production motive. The emerging exploitive farming community in India, comprising army personnel, business men and the educated urban public; should become aware of this. Intensive cultivation of land without conservation of soil fertility and soil structure would lead ultimately to springing up of deserts. Irrigation without arrangements for drainage would result in soils getting alkaline or saline. Indiscriminate use of pesticides, fungicides and herbicides could cause adverse changes in biological balance as well as lead to an increase in the incidence of cancer and other diseases, through the toxic residues being present in the grains and other edible parts. Unscientific tapping of the underground water would lead to rapid exhaustion of this wonderful capital resource left to us through ages of natural farming. The rapid replacement of numerous locally adapted varieties with 1 or 2 high yielding strains in large contiguous areas would result in the spread of serious diseases capable of wiping out entire crops, as happened prior to the Irish potato famine in 1845 and the Bengal rice famine of 1942. Therefore, initiation of exploitive agriculture without a proper understanding of the various consequences of every one of the changes introduced into a traditional agriculture and without first building up a proper scientific and training base to sustain it, may only lead us into an era of agricultural disaster in the long run, rather than to an era of agricultural prosperity.

In the same Presidential Address that will remain for a long time in the annals of Indian agriculture, Swaminathan also referred to protein and caloric hunger. I cannot really think of any other cytogeneticist, and an authority on radiation and chemical mutagenesis making a quantum jump into an entirely different area of science and direct the developments for quenching hunger — both protein and caloric. Swaminathan referred to a survey by the UN Food and Agricultural Organisation (FAO) and pointed out that countries with the lowest *per capita* daily protein and caloric consumption are also those with the lowest productivity. He cautioned, *Unless the widespread malnutrition-induced physical inertia can be banished soon, social conditions conducive to technological advance will be hard to create.*

6.3 The Impact of Dwarfing Genes on Wheat and Rice Production

At the outset, Swaminathan [126] explains the origin of the semi-dwarf wheats.

A Japanese semi-dwarf wheat variety *Daruma* was probably the starting material. It is crossed with the variety *Fultz* (a US winter wheat, high-yielding). This cross resulted in a semi-dwarf, high-yielding variety of the *Fultz–Daruma*. The *Fultz–Daruma* was crossed with *Turkey Red* (US winter, high-yielding) and *Norin-10* resulted from this cross. *Norin-10* (semi-dwarf, high-yielding winter wheat) was crossed with local winter varieties adapted to US North West and *Gaines* (semi-dwarf, US adapted winter wheat) resulted. *Gaines* was crossed with local strains to develop *new wheats* (semi-dwarf, high-yield, adaptable, rust-resistant, fast maturing, spring). Figure 6.2 explains what is given in the text.

Daruma   **X**   Fultz
(Japanese semi-dwarf)   ↓   (US winter wheat, high-yield)

Fultz–Daruma   **X**   Turkey Red
(semi-dwarf, high-yield)   ↓   (US winter, high-yield)

Locals   **X**   Norin-10
(adapted to US North West)   ↓   (semi-dwarf, winter, high-yield)

Gaines   **X**   Local Strains
(semi-dwarf, winter, US adapted)   ↓

New Wheats
(semi-dwarf, high-yield, adaptable,
rust-resistant, fast-maturing, spring)

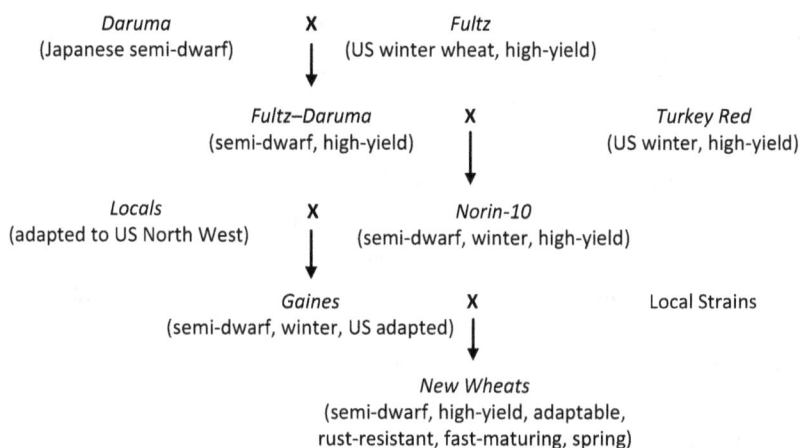

Figure 6.2. Origin of semi-dwarf wheats [126].

The semi-dwarf and dwarf spring wheat varieties were developed in Mexico by Dr. N.E. Borlaug. In order to develop dwarf wheat varieties suitable for cultivation in India, the IARI, New Delhi introduced in 1963, at Swaminathan's instance, a large variety of wheat material containing the *Norin* dwarfing genes from Mexico through the courtesy of the Rockefeller Foundation and the Mexican Ministry of Agriculture.

Swaminathan [126] has stated that even in the first year, i.e. 1963, the IARI distributed this material to several wheat breeding centres in order to assess the extent of their adaptation and reactions to the races of black, brown and yellow rusts prevalent in the country. In addition to breeding material, bulk quantities of four commercial spring wheat varieties, *Lerma Rojo 64A, Sonora 63, Sonora 64 and Mayo 64* were also obtained. These varieties have been tested in all the wheat-growing states of India during four 'rabi' seasons under the All India Coordinated Wheat Improvement Project. In addition, they were subjected to detailed physiological, pathological, chemical and agronomic tests at the IARI. Two of these varieties, *Lerma Rojo 64A and Sonora 64*, were approved by the Central Variety Release Committee of the Indian Council of Agricultural Research (ICAR) in 1965 for cultivation in the irrigated areas. These provided the initial material for the High-Yield Varieties Programme started in 1967. *Lerma*

Rojo is a late variety with a high degree of resistance to yellow rust. It performed well under timely sowing conditions and in areas characterised by yellow rust epidemics. That *Sonora 64* is an early variety and is well-suited for being grown in rotations has already been discussed.

Swaminathan also wrote a great deal about dwarf rice varieties. The genes conferring a short plant stature have also had a profound impact in increasing the yield potential of rice. Many of the *indica* varieties developed in Taiwan and other places have a short, non-lodging habit. The *Taichung Native No. 1*, which yielded more than 6,000 kg/ha in some parts of India, had in its pedigree the recessive dwarfing gene which occurred as a spontaneous mutation in the variety *Dee-Gee-Woo-Gen*. This dwarf (80 cm in height but with good tillering and high-yielding capacity) was crossed with *Tsai-Yeun-Chung* (130 cm in height) and *Taichung Native No. 1* was developed by selection in the segregating population.

6.4 Christening the High Level Fertiliser-Responsive Dwarf Wheat and Rice Varieties

As already described, Swaminathan [122] referred to the new agriculture to bring about dramatic yield increases based on high level inputs of chemical fertilisers, chemical pesticides and copious irrigation and contiguous cultivation of one or two high-yielding wheat and rice varieties without concomitant cultivation, conservation and consumption of numerous locally adapted varieties as *exploitative agriculture*, that in long time would result in agricultural disaster than agricultural prosperity. In fact, he emphasised the negative aspects of the *new agriculture* in clear and precise terms. The Government of India special 'Wheat Revolution Commemoration Stamp' was released after about 7 months (July 17, 1968) of Swaminathan's epoch-making Presidential Address (Fig. 6.3).

Swaminathan's view has rightly been that *exploitative agriculture* was absolutely essential to prime a system of intensive agriculture to break the centuries-old stagnation of yields/productivity. But then that system should be used only to get some 'breathing space'. Then, a more sustainable-meaning eco-friendly, as well as socially equitable system of agriculture should replace the *exploitative agriculture*. However, William Gadd of the US Agency for International Development

Figure 6.3. **Wheat revolution commemoration stamp 1968.**

christened the exploitative agriculture as the 'Green Revolution'. The Green Revolution was variously interpreted to mean a biological or a political event. The term Green Revolution could, in one sense, mean that India's wheat yields had remained static for over hundreds of years, and the dwarf and semi-dwarf wheats had brought about quantum jump in the increment of the yield. Swaminathan had pointed out in several of his lectures that India's progress in wheat production in just 4 years (1964–1968) was equal, if not substantially more than that during the preceding 4,000 years. From a basic biological point of view, the colour of the *chlorophyll*, the pigment involved in the photosynthesis is green and the grain production (yield) involves efficient harvesting of the appropriate wavelength (i.e. energy) of the sunlight. In an extreme deviation from the biological interpretations is the political sarcasm, to imply that Green Revolution symbolising food security and peace is just the opposite of a violent Red Revolution that established the former Soviet regime.

Swaminathan accepted the term Green Revolution and in fact, coined the term 'Green Revolution Symphony' because of the mutually reinforcing packages of technology, services, public policies and farmers' enthusiasm.

His according *full credit to the farmers for the yield revolution directly links the product of science with society.* Writing in the *Illustrated Weekly of India* Swaminathan [130] made the following remarks on the Punjab Wheat Miracle:

> *Brimming with enthusiasm, hard-working, skilled and determined, the Punjab farmers have been the backbone of the revolution. Revolutions are usually associated with the young, but in this revolution, age has been no obstacle in the participation. Farmers, young and old, educated and uneducated have easily taken to new agronomy. It has been heart-warming to see the young college students, retired officials, ex-army men, illiterate peasants and small farmers queuing up to get the new seeds. At least in the Punjab, the divorce between intellect and labour which has been the bone of our agriculture is vanishing.*

There have been many positive and negative impacts on environment of Green Revolution. The positive aspects have been the following:

Because of increased productivity, the land required to produce a certain quantity of cereal grains had been contained within certain limits. Had the productivity not substantially increased, more forest land would have been required to be brought under cultivation (i.e. conversion of forest land into farmland). Hence, Swaminathan referred to it as 'land-saving' or 'forest-saving' agriculture. However, it was Norman Borlaug who continually advocated increasing crop yields, as a means *to curb deforestation.* The view that *increasing the productivity of agriculture on the best farmland can help control deforestation by reducing demand for new farmland is referred to as Borlaug hypothesis.* Now there are growing scientific arguments against the 'Borlaug hypothesis' both in terms of the basic assumption as well as the quantity of forestland so far saved. One of these arguments is that the deforestation does not occur only for the farming purposes; in fact, housing, industries, roads, new townships usurp substantially larger areas of the forestland. Stevenson *et al.* [131] point out that their results generally support the 'Borlaug hypothesis', but this relationship (i.e. between increases in yield of cereal crops and saving of forest land) is quite complex, and the net effect is of a much smaller magnitude than Borlaug proposed. They estimate that the total crop area in 2004 would have been between 17.9 and 26.7 million hectares larger in a world that had not benefitted from crop germplasm improvement since 1965. Of these hectares, 12.0–17.7 million would have been in developing countries, displacing pastures and resulting in an estimated two million hectares of additional deforestation.

The magnificent effect of the Green Revolution in India was the transformation of its then image as a 'begging bowl' to 'bread basket'. It was indeed an answer to the Malthusian doom predicted for India by Ehrlich [124], Paddock brothers [125] in their books discussed earlier. The social dimension of the Green Revolution too has both positive and negative dimensions. On the positive side, it rejuvenated the self-confidence and self-reliance of the farmers in general, and the Punjab farmers in particular. In this regard, Swaminathan's article [130] has already been referred to. Significantly, the annual rate of increase in the cereal grain production had been put ahead of the annual population growth rate. The dramatic improvement in the national agricultural scenario stimulated all round developments in industries, defence, corporate sector and national employment and services sector. India was able to conduct the first peaceful nuclear explosion test in 1974. Subsequently, Swaminathan observed that the Green Revolution-led National food security provided the necessary political freedom to conduct such nuclear tests. Had India been still dependent on food imports to feed her people, it might not have conducted such tests owing to fear of unpleasant and unmanageable consequences such as ban on food export to India. So, food security is necessary for political sovereignty and the Green Revolution secured that to India. The negative dimensions of the Green Revolution justify the reference to it as *exploitative agriculture*. Although clearly foreseen and forewarned by Swaminathan as early as 1968, nothing could prevent this high-yielding, commodity-centric *exploitative agriculture* from progressively degenerating into a *Greed Revolution*; the point is that 'greed' is a very powerful and insatiable trait of the humans, but towards self-annihilation in the long term. The infamous rice–wheat rotation without leguminous crops in the crop rotation was notably a harmful agricultural practice. The provision of free electricity led to flooding the soil with groundwater, and without proper drainage the soil soon became alkaline. Excessive use of chemical fertilisers and pesticides adversely affected the beneficial soil microorganisms and earthworms. The chemical pesticides left carcinogenic residue on the edible parts of the plants. Today, it is agreed that the Green Revolution belt has a large number of farmers, members of the farming families as well as general public in the farming villages afflicted with one or the other form of cancer. As was also fore-warned, many of the locally-adapted excellent local wheat varieties have become extinct due to their not being in the cultivation, conservation, con-sumption and commercialisation (the 4Cs propounded by Swaminathan) chain.

Yet another spin-off of the Green Revolution was that it only enhanced the food *availability* in the country and established the food security at the national level, but not at the individual household level of millions of rural poor. The negative dimensions of the Green Revolution started showing up by the mid-1980s. Swaminathan's reference to the 'new high-yielding agriculture' as *exploitative agriculture* as early as January 1968 proved right. There is no evidence to suggest that Swaminathan had foreseen a rice–wheat–rice rotation in the best soils of Punjab when he made his Presidential Address to the Agricultural Section of the *55th Indian Science Congress Session* in January 1968 at Varanasi [122]. The point is that a subsequent adoption of an unscientific rice–wheat–rice rotation, and its progressive intensification accelerated the pace of degeneration of the 'Green Revolution' into a 'Greed Revolution'. Swaminathan has done a very fair job of analysing the factors involved and openly stating the things to come under a set of given circumstances. He could not have done more than that in an intensely democratic coun-try in which the politicians prefer to strengthen their 'vote banks', than prevent such decisions and actions that would go against their ambitions; the bulk of the people of the largest democracy in the world know full well that political expediency in several sectors gain priority over the social, gender and environmental problems of the country. It is now established beyond any doubt that the intensification of the rice–wheat–rice cultivation *without* leguminous crops in crop rotation has been one of the major factors for the degeneration of the Green Revolution into Greed Revolution. Extensive research studies suggest that millions of tonnes of cereal grains stored in the warehouses go rotting, but the poor are unable to purchase them and would remain hungry. In the 1980s, articles appeared in the print media that the country was witnessing a paradox, *mountains of grains on one hand, and millions of hungry people on the other.* In fact, *Business Week* (November 07, 1994) wrote: *Even though the Indian granaries are overflowing now, 5,000 children die each day of malnutrition, and one-third of India's 900 million people are now poverty-stricken. This explains the fact that in South Asia where the Green Revolution seeds have contributed to the greatest production success, roughly two-thirds of the undernourished in the entire world live. It should, however, be noted that during the two decades from 1970 to 1990, the total food available per person in the world rose by 11 percent, while the estimated number of hungry people fell from 942 million to 786, a 16% drop.* Since then (1990), the absolute number of hungry people has been gradually increasing and it is now around 1 billion people out of the total global population of about 7.4 billion. India is home to about 300 million of them.

The cause of hunger in South Asia, particularly India, is because of lack of money that provides purchasing power. In his book [132], Swaminathan has cited Colonel Baird Smith, who as early as 1856 investigated the possible causes of a serious famine in Northwest India and wrote: *Indian famines are famines of work and not of food. Where there is work, there is money. Where there is money, there is food.* About 6 years later, in 1862, he wrote: *India we all know very well that when agricultural class is weak, the weakness of all other sections of the community is the inevitable consequence.* Swaminathan points out that the Green Revolution was not, however, designed to produce rural livelihoods; hence the number of hungry people grew almost concomitant with the growth in human population.

6.5 Swaminathan Said It in 1968 and Others Said Two Decades Later

Swaminathan's epoch-making Address in January 1968 [122] has already been discussed. While it will remain in the annals of the global agriculture in general, and Indian agriculture in particular, the unfortunate part of it has been that no remedial actions were implemented in time to check the harm to ecology and social/gender equities. Consequently, the Green Revolution resulted in considerable damage to the ecological and social pillars of sustainable agriculture. The proof of this comes from several recent research papers, and only two of these are cited here to substantiate that what Swaminathan had forewarned in 1968. An interesting paper is by Bourne [133], an Award-winning journalist who wrote a Special Report entitled *The End of Plenty*. Referring to the fragile food web, he says, *the global food crisis did not happen overnight and that the challenge today is not to deal with short-term rise in grain prices, but to find ways to avoid perpetual food crisis.* He goes on: *The Green Revolution that Borlaug started had nothing to do with the eco-friendly green label in vogue today. With its use of synthetic fertilisers and pesticides to nurture vast fields of the same crop, a practice known as monoculture, this new method of industrial farming was antithesis of today's organic trend ... Today, though, the miracle of Green Revolution is over in Punjab; yield growth has essentially flattened since the mid-1990s. Over-irrigation has led to steep drop in the water table, now tapped by 1.3 million tube-wells, while thousands of productive land has been lost to salinization and water-logged soils. Forty years of intensive irrigation, fertilization, and pesticides have not been kind to the loamy grey fields of Punjab. Nor, in some cases to the people themselves.* Bourne's [133] last sentence refers to the increased incidence of all kinds of cancer to the members of the farming families as also to others due to recurrent exposure to pesticides and their residues. A word by word comparison of what Swaminathan wrote in 1968 and what Bourne [133] wrote four decades later in 2009 reveals the science-based prophecy made four decades ago, now stands proven to be compatible with the ground reality in the long-term.

In an article on the national food security *vis-à-vis* sustainability of agriculture in high productivity regions, Dhillon *et al.* [134] have presented a wealth of data which reveals a clear plateauing of the productivity of wheat and rice in Punjab since 1996–1997. The trend has continued through 2008–2009 when the paper was written. Further, the authors point out the intensive rice–wheat cultivation had created a serious stress on soil and water resources and the very sustainability of crop production is threatened. The cultivation of fodder and forage as well as grain legumes had been left out. I agree with the view expressed a long time ago by Swaminathan [122] that the Green Revolution would degenerate into a Greed Revolution; this has since become true.

6.6 A Lesson for the Younger Generation of Scientists and Technologists

Given the normal human nature, there is no doubt that Swaminathan did something very extraordinary that most others would not have. He sought to address the possible long-term detrimental consequences of the new agriculture that was obviously harsh on the ecological foundations of sustainable agriculture (i.e. soil health, freshwater, biodiversity, clean and renewable energy which fossil fuel-based farm operations as well as the inorganic nitrogen fertiliser produced by high level energy-based synthesis of nitrogen and hydrogen by the Haber–Bosch Process) and atmosphere (mitigation of carbon to retard the climate change) as well as social and gender inequities. From these points of view, it is evident that Norman Borlaug did not view these issues as Swaminathan did. The younger scientists ought to get it imprinted in their hearts, that scientific brilliance and achievements alone, without honesty and concern for both the society and ecology would inevitably bring disaster to both themselves and the people for whom they practice science. The honesty and integrity of Swaminathan indeed make him a 'legend in science and beyond'.

Any book written for posterity should allude to the facts as they have been at a point of time given in the history. Hence, a clarification is made here to an analysis made by Ruth DeFries [135] in her book, *The Big Ratchet: How Humanity Thrives in the Face of Natural Crisis*. Ruth DeFries describes the environmental and socio-economic problems associated with the Green Revolution. She brings up for discussion some 'perceived differences' in views between Swaminathan and Borlaug on the Green Revolution. She is right in writing that achieving dramatic increase in the crop productivity in a short period at the national level was indeed a major 'pivot', and once the 'pivot' pushed the human capability forward and to a higher level, a 'ratchet' (i.e. a wheel with a rim so toothed as to move in one direction only) locks the gains brought about by the pivot. So, the argument is that Borlaug's Green Revolution should be taken to regions not yet covered by it, and further intensified. She writes: *Swaminathan does not see it that way, instead, he has been proposing something called 'Evergreen Revolution' and vigorously campaigns for the transformation of the Green Revolution into an Evergreen Revolution.* Further, Ruth makes a statement in her book, *whether the future veers toward Swaminathan's or Borlaug's view, one message is clear; the Green Revolution was yet another experiment in feeding the humanity.* In review of Ruth's book [136], I have elaborated how a technology engineered to provide only a 'breathing space' (as Swaminathan had put it) could not continue as a 'pivot' (to elevate humankind to food security for all times to come) and that it has degenerated into a 'hatchet', (that destroys the very production base). There has been growing evidence that the Green Revolution was gradually becoming a hatchet since the 1990s [136]. Ruth did not cite Wilson [137], a biologist of rare distinction who observes in his epoch-making book, *The future of life* that Swaminathan's Evergreen Revolution is the best option available to feed the burgeoning human population and save the rest of life as well, since it is 'pivoted' on a systems approach. Wilson wrote: *The problem before us is to feed billions of new mouths over the next several decades and save the rest of life at the same time, without being trapped in a Faustian bargain that threatens freedom and security. No one knows the exact solution to the dilemma. Most scientists and economists who have studied both sides of it agree that benefits outweigh risks. The benefits must come from Evergreen Revolution.* He has mentioned Swaminathan by name in his famous book.

There is yet another major global decision in favour of sustainable development in every sphere of human activity including agriculture, food and nutrition security and poverty reduction. By the end of 2015, the tenure of the 'Millennium Development Goals (MDGs)' (2000–2015) with its 8 Goals and 18 targets has ended. The new set of goals and targets aims at greater inclusion of the environmental and social dimensions of the sustainable development. The MDGs greatly focused on the economic than on the environmental (ecological) and social components. Hence, the 'sustainable development goals' (SDGs) replace the MDGs for the next 15 years (2016–2030). While more about these would follow in later chapters, the emphasis here is on the fact that Evergreen Revolution very naturally fits in well SDGs and the 'green economy', because of its 'systems approach'. The world needs ideas to help planet Earth and humanity to slow down the pace of onset of 'tipping point' with regard to climate change and breakdown (collapse) of civilisations. It is not a question of 'who said it', but of 'what has been said and whether the prescription will help toward sustainable development or not'.

The people in general and the youth aspiring to be scientists and technologists, in particular, would do well to emulate Swaminathan for dispassionately evaluating their contributions for its long-term impact. It is emphasised that Swaminathan did not seek glorification for his having brought about the Green Revolution although a grateful nation and its people accorded notable recognition — yet he was not awarded the 'Bharat Ratna' for his changing the image of the nation from 'begging bowl' to 'bread basket'. In that omission or denial, the nation and its leadership over the past four decades seem to have forgotten that 'food to quench hunger' is the most basic human need in a hierarchy of needs, and far more important for survival and peace than one's accomplishment in any other area. People can live without anything else but food. Those who have known the hard journey he has made, and is still making, to create a 'hunger-free world' shall definitely place on record that Indian political leadership since 1970s has failed to accord due recognition to Professor M.S. Swaminathan whose achievements coupled with his contributions to humanity and planet Earth have hardly any parallel. In fact, achieving food security at the national level is possibly the most noteworthy achievement in the 20th century India, after its independence.

6.7 India's Hunger Problem — Availability of Food Alone Will Not Do

While India's agriculture is a gamble with both rain (monsoon) and temperature during the grain formation stage, the food security at the individual household level is a gamble with livelihood opportunities, especially for about 650 million women and men living in about 6,38,000 villages of 'Bharat', that is India. Earlier, it has been described that granaries overflowing with cereal grains in India have not eliminated the hunger of over 300 million people. The remarkable aspect of Swaminathan's approach to ensure food security at the individual household level of hundreds of millions of people in India is to integrate food production and thus making 'food availability' with building a broad rural livelihood base to generate income to access food. I am not aware of anyone else anywhere in the world to tackle the twin problems in an integrated manner concurrently.

The year was 1972 — Swaminathan had just taken over as the Director-General of the ICAR, and Secretary to the Government of India, Department of Agricultural Research and Education in the Union Ministry of Agriculture. Almost immediately after that, he was invited to deliver, 'The Princess Leelavathi Memorial Lecture' on January 17, 1972, at the University of Mysore. The title of his lecture was, *Agricultural Evolution, Productive Employment and Rural Prosperity* [138].

There are several golden statements which, in my studied view, are the basic elements of peace and harmony, conservation and cultivation, and technology and ecology. Therefore, I would like to cite some of these in the words of Swaminathan.

1. *The problem of a technological upgrading of agriculture is a complex one involving a dynamic interaction of a scientific and socio-economic factors with the environment.* Over five decades ago, he noted that technology needs to be both eco-friendly and socially equitable. Few scientists have ever said so.

2. *Half of a scientist's difficulties are over once the problem needing solution is clearly formulated.*

3. *Development of varieties suitable for increasing the income and employment potential of farms of small size through a series of multiple and mixed cropping systems, based on choices and alternatives from an economic standpoint and on sound principles of ecology from the scientific point of view.*

4. *Collection and conservation of gene pool.*

5. *An important aspect of the agronomic practices should be the 'breeding' of soils suited for sustained high productivity. This would involve simultaneous attention to the physical, biological, chemical and topographical facets of soil fertility. Intensive agriculture based on modern plant breeding without coincident steps in soil breeding would result ultimately in agricultural disaster rather than agricultural progress. Unfortunately, there has been little realization of this position so far.*

6. *Development of an Ecology-cum-Economics based Multiple-Cropping 'Cafetaria' system. Multiple or relay cropping would help to increase*

the income and employment potential of holdings of small size. However, intensive exploitation of land and unscientific crop rotations could lead to very undesirable long term consequences. There should, therefore, be some ground rules in the introduction of multiple cropping patterns. Among these are:

- *No two crops sharing in common the same pests should be grown in succession.*
- *One crop should be deep rooted and another with more shallow roots, so that different layers of the soil can be tapped for nutrients.*
- *For the restoration of soil fertility, at least one leguminous crop should find a place in the rotation. If the stalks of maize are not needed for feeding cattle, their incorporation in the soil would help to improve soil structure. On the above lines, we could evolve for each agro-climatic area, a multiple cropping 'Cafetaria' with a wide choice of alternative crops from which the farmer can choose an appropriate combination suited to his need and input — mobilising potential and the demands of the market.*
- *Multiple cropping is a very potent instrument for improving the economic position of a farmer with a small holding and for banishing under-employment and unemployment.*
- *With the spread of the new technology and the dramatic transformation of agriculture in certain areas, new dimensions of adult education also appear. There are new needs for education among farming communities. There is a great hunger not only for new knowledge related to agriculture, but also for new skills, particularly technical skill connected with it. The demand for 'techniracy' if one may coin a term based on the idea of technical literacy, is likely to be stronger and deeper and also more widespread than for formal literacy, or even for functional literacy. New approaches to adult education must capitalize on this new demand and need for 'techniracy'.*

Swaminathan's concept and description of what could be the areas under 'techniracy' for the largely illiterate and unskilled rural men and women are of practical value in diminishing the rural poverty.

He went further to suggest short residential courses in 'tech-niracy'. These include topics like tube-well construction, choice of pumps, drip irrigation, integrated pest and nutrition management, etc. The list he prepared in 1972 could be substantially expanded with the kind of new challenges (e.g. climate change, loss of bio-diversity, soil health degradation, etc.) and the eco-technological (i.e. technologies resulting from blending frontier technologies such as nuclear, space, and information and communication tech-nologies (ICTs) with ecological prudence and traditional knowl-edge of the rural and tribal people, particularly women) solutions to solve/overcome the problems.

- Swaminathan also suggested ways and means of bypassing the literacy problem. He stated: *To the extent that mass media can succeed in (a) transmitting the necessary information and (b) dem-onstrating frequently the necessary skills for daily living and suc-cessful farming, the question of formal literacy can be bypassed. Radio and films have to be used more intensively, and of course, television in those areas where it is already or will soon be availa-ble. This means that the specifically technical programmes should be still more specific and geared to the needs and frequently repeated. It also requires a specific information or skills but more closely related to rural life in all its aspects. In the audio-visual world of the future, formal literacy may not be necessary for the vast majority, but only for those interested in pursuing education at higher levels. This is likely to be true also for the nations at the other end of the technological spectrum, for different reasons. Books may soon be getting out of date. Hence, if we can make the jump from the pre-literate to the post-literate world, without passing through the intermediate phase, we can save enormous resources and time. Planned expansion and improvement of the mass media would hence make a major contribution to the diffusion and correct adoption of new technology.*

- *Challenge and opportunity: At no time in human history, have so few had so much opportunity for service to fellow-men, as the edu-cated classes have today in our country. If we rise to the challenge,*

*ours can still be the land where science and spirituality become
blended to create a truly happy land.*

The extensive quoting of Swaminathan in his own words is indeed the
most effective way in which his vision, thoughts, words and intended
actions at the appropriate time could be presented so that the readers could
go beyond just reading to absorb and analyse the contemporary relevance
of what he had said over five decades ago.

I believe that 'The Princess Leelavathi Memorial Lecture' on January
17, 1972 by Swaminathan [138] has had several indications and guidelines about the programme and their inter-linkages at the M.S. Swaminathan
Research Foundation (MSSRF) he was to set up within the next two
decades. In fact, he set up the MSSRF, his 'Magnum Opus', in 1988. As
one could see, the major statements made by Swaminathan (reproduced in
bullet points) bring out (i) how both agriculture and technology need to be
eco-friendly, (ii) socially relevant, (iii) capable of generating rural livelihoods and (iv) how the rapid and breath-taking advances in the ICT could
help in agriculture and rural development and (v) how the ICT coupled
with 'techniracy' would greatly help in the knowledge and skill empowerment of the rural women and men who have not had the formal education
and academic literacy.

6.8 Going Beyond Caloric Hunger-Combating Malnutrition (*Hidden Hunger*)

Swaminathan's vigorous campaign against malnutrition integrated with
'caloric' and 'protein' hunger since the last decade might give an erroneous impression to many people that he has begun to think about hidden
hunger only in the recent times. That impression, if any, is wrong. As early
as 1973, writing on *Malnutrition in a Two-Tier World* [139], he made an
important observation: *The paradox of nutrition in this compartmented
world (i.e. rich vs poor countries) is that malnutrition among the poor is
increasing at the same time as the realisation of the importance of nutrition. Nutrition and education are now often compared to the hardware
and software components of development. It is unfortunate, but true, that*

there can be little use for the software, unless hardware component exists. Evidence is now accumulating and suggests that the protein–caloric malnutrition in early life has lasting effects on the anatomy, biochemistry and functioning of the central nervous system.

6.9 Managing Widespread Drought Without Food Imports

It was on the occasion of the 'Dr. Rajendra Prasad Memorial Lecture' delivered on March 26, 1972 at the Indian Statistics Society, New Delhi, Swaminathan titled his talk; *Can we face a widespread Drought Again without Food Imports?* He had put the 1965–1966 drought and the drop in India's food production by about 19%, which amounted to about 17 million tons in quantitative terms, as the backdrop to his Address. Import of food mainly from the US under PL-480 helped in averting a widespread famine. His prescription was to form an organisation of a National Crop Planning Board charged with the responsibility of developing alternative cropping strategies for every part of the country on the basis of different weather models. There should be contingency plans ready to meet crop failures arising from drought or floods, so as to ensure that not only does total crop production remain fairly normal under such conditions but also that the farmer has reasonable income and livelihood. In addition to buffer stocks of grains, there should also be buffer stocks of seeds of drought or flood-avoiding and photo-insensitive and quick yielding varieties of crops. He also suggested that we be ready with plans for optimal use of water, fertiliser and pesticides during drought years.

Chapter 7

Ushering in the Evergreen Revolution

7.1 Eco-Agriculture

The earlier chapters brought out at least three major concerns frequently discussed by Swaminathan in public lectures and publications, comprise the following:

(i) Green Revolution would cause serious harm to environment, including the ecological foundations of sustainable agriculture, further deepen social and gender inequities and would not ensure food and nutrition security at the individual household level of hundreds of millions of resource-poor farming, fishing and landless rural families. For these reasons, it would not be sustainable in the long run; in fact, it could lead to an agrarian crisis, unless it is prevented from becoming a 'Greed revolution'. By the late 1980s, field-based data by several scientists revealed that yield gains associated with the Green Revolution had already begun to suffer a fatigue due to degradation of soil, fresh water quality, etc.

(ii) In order to ensure food security at the individual household level, food availability alone will not do; concomitant with food production, the resource base in rural areas for enhancing on-farm and non-farm livelihoods for income generation and thus 'access' to food should not only be conserved but also enhanced. Access to these resources should have social and gender equities.

(iii) Food security in terms of providing caloric needs alone will not provide 'balanced diet' and good health for productive life. Besides the calories provided by the consumption of cereal grains, it is essential to include proteins (e.g. pulses, fish, meat, etc.) and micronutrients (fruits and vegetables especially those naturally fortified with iron, iodine, zinc, vitamins, etc.) in the diet. Therefore, a system of agriculture that is completely free from the negative aspects of the Green Revolution has become an essential need.

While Swaminathan [122] had clearly put on record as early as 1968 the negative dimensions of the Green Revolution, the need to fight poverty, malnutrition and environmental degradation was focused in 1972 at the United Nations Conference on *Human Environment* in Stockholm, Sweden. It was intended to awaken the conscience of citizens of the world and national leaders about environmental degradation and its repercussions on human well-being. This conference would have ignored the poverty and hunger which constitute the social dimensions of conservation of the resources, but for a very powerful statement stressing the undeniable link between poverty and environmental degradation made by the then Prime Minister of India, Smt. Indira Gandhi. Leading the Indian Delegation to the conference she spoke: *Are not poverty and need the greatest polluters? For instance, unless we are in a position to provide employment and purchasing power for the daily necessities of the tribal people, and those who live in and around jungles, we cannot prevent them from combing the forest for food and livelihood; from poaching and from despoiling the vegetation. When they themselves feel deprived, how can we urge the preservation of animals? How can we speak to those who live in villages and in slums about keeping the oceans, the rivers and the air clean when their own lives are contaminated at the source? The environment cannot be improved in conditions of poverty. Nor can poverty be eradicated without the use of science and technology.*

Those who have been closely following Swaminathan's lectures, and writings since 1968 might note that the then Prime Minister Indira Gandhi had indeed spoken all that Swaminathan had been talking since 1968. She has not, of course, openly referred to food and nutrition security but none can dispute the fact that abject poverty is the root cause of hunger.

The major outcome of the UN Conference on *Human Environment* in 1972 was the consensus that poverty-led hunger and environmental degradation form a vicious spiral with one accentuating the other. This greatly strengthened Swaminathan's resolve to develop/design eco-friendly agriculture integrated with eco-technologies to enhance the rural livelihoods. He started developing a blueprint while he was extremely occupied in the 1970s and until the later part of 1980s, with greatly responsible and highly-demanding positions as the Director, Indian Agricultural Research Institute (IARI), Director-General, Indian Council of Agricultural Research (ICAR), Member Planning Commission, Government of India, Director-General, International Rice Research Institute (IRRI), Los Banos, The Philippines, etc., in addition to several International assignments. He was able to implement his concepts of sustainable agriculture, and rural development (i.e. Evergreen Revolution) only after he set up his 'Magnum Opus', the 'M.S. Swaminathan Research Foundation' (MSSRF) at Chennai in 1988–1989.

In the years following the UN Conference on Human Environment in Stockholm in 1972, the major concern was to reconcile development and conservation of environment. The consensus was for sustainable development, which was, if at all, only vaguely comprehended. In fact, many wondered if it was an oxymoron. It was therefore, timely that the World Commission on Environment and Development (WCED) was set up by the United Nations with the mandate to come up with a report on sustainable development. Mrs. Gro Harlem Brundtland former Prime Minister of Norway was appointed the Chairperson. Her report [140] entitled *Our Common Future* among other things, defined sustainable development as *development that meets the needs of the present without compromising the ability of future generations to meet their own needs.* More than the definition of the sustainable development, its title has greater significance. It is so because it emphasises that the human destiny lies with that of the planet, and in fact, it states in clear and simple words that notwithstanding political and geographical barriers, the life of all humans on this plant is ecologically entwined.

There could be no fruitful discussion on sustainable development 'without' agriculture and food security being the most basic issue. Therefore, the WCED had also set up the Panel on 'Food Security, Agriculture, Forestry and Environment' and asked the Panel to suggest ways in which humankind could be insulated from hunger on an ecologically sustainable basis. Professor M.S. Swaminathan was appointed the Chairman, Advisory Panel on Food Security. The Swaminathan Panel proposed a Seven-point Action Plan consisting largely of macro-solutions to the Agriculture–Environment problems of global dimensions. The Swaminathan Panel Report [141] to the WCED was given the title, *Food 2000 — Global Policies for Sustainable Agriculture* first published in 1987. The beginning of Chapter 1, entitled *The Challenge* reveals the level of unsustainable production and also touches on the mountains of grain stocks on one side, and 18 million people dying of hunger each year in the 1980s. Swaminathan Panel describes the appalling situation as a cruel paradox. The Panel minced no words in stating, *The mountains of surpluses are a result of interrelated and ecologically unsound agricultural, economic, trade and aid policies.*

The Swaminathan Panel defined the term 'Sustainable livelihood security' as follows:

(i) Livelihood is defined as adequate stocks and flows of food and cash to meet basic needs.
(ii) Security refers to secure ownership of, or access to resources and income-earning activities, including reserves and assets to offset risk, ease shocks and meet contingencies.
(iii) The term 'sustainable' refers to the maintenance or enhancement of resource productivity on a long-term basis.

The Panel elaborated on livelihood security as basic to achieving a stable human population, good husbandry, sustainable management and effective means to reverse destabilising processes. Swaminathan introduced in the Report reference to Mahatma Gandhi and his approach to development of the *Antyodaya* model to emphasise that priorities in development should be measured by their potential benefit to the weakest and the poorest sections of the community. The Seven-point Action Plan of Swaminathan Panel [141] to achieve sustainable agriculture:

1. developing an international code for sustainable and equitable use of life-support systems,
2. including sustainable livelihood for all, in the UN Declaration on Human Rights,
3. initiating a new agricultural system for nutrition security,
4. ensuring equality of opportunity for access to technology,
5. organising skills for sustainable livelihood security in every country,
6. reorienting international action and assistance so that it is to be consistent with integrated national conservation and sustainable livelihood strategies, and
7. promoting political commitment and accountability.

The emphasis on sustainable livelihood security is of foremost importance. The Panel Report [141] in its page 6 points out that as of then, *almost 20% of the world's population who suffer from starvation and malnutrition do so, not because there is shortage of food, but because they do not possess the income needed to command their share. Indeed, 50% of the world's hungry people live in just five countries, four of which are in Asia where the Green Revolution has taken place and even in one such country where national surpluses have been recorded.*

The Swaminathan Panel Report to the WCED formed the basis of Gro Harlem Brundtland's reference to sustainable agriculture and livelihood security in her Report *Our Common Future* [140].

Swaminathan's transformation of the Green Revolution into sustainable Evergreen Revolution greatly reflected the recommendations of the WCED Panel for sustainable agriculture of which he was the Chairman. In this regard, reference is made to a few of his books he had written in the 1990s; these elaborate the then prevalent conditions of natural resources, progressive degradation of the ecological foundations of agriculture, burgeoning human population and growing social and gender inequities. These also touch upon the imminent onset of climate change. His earlier books on the concept of Evergreen Revolution and the necessity to transform the Green Revolution into a sustainable *Evergreen Revolution* are as follows:

(i) *Sustainable Agriculture: Towards an Evergreen Revolution* [142];
(ii) *Sustainable Agriculture: Towards Food Security* [143];
(iii) *I Predict: A Century of Hope. Towards an Era of Harmony with Nature and Freedom from Hunger* [144].

Swaminathan's article *An Evergreen Revolution* [145] defined his new concept of sustainable agriculture (Evergreen Revolution) *as achieving productivity in perpetuity 'without' accompanying ecological and social harm.* Two years later in 2002, Swaminathan's book [132] *From Rio de Janeiro to Johannesburg — Action Today and Not Just Promises for Tomorrow* was published. In the Introduction to Chapter 1 (Sustainable Food and Water Security) Swaminathan [132] writes: *It is widely agreed that the breathing spell provided by the Green Revolution for achieving a balance between population growth and food production will soon get exhausted, unless we take steps to foster an Evergreen Revolution based on principles of ecology, gender and social equity, economics, employment generation and energy conservation.*

He had thus made it abundantly clear, sustaining productivity in perpetuity and enabling 'access' to food at the individual household level especially of the resource-poor marginal farmers, fishing and landless families in the rural areas involved concurrent attention to several complex and inter-related factors from the domains of ecology, social and gender issues, economics, eco-technologies as well as skill and knowledge empowerment of the people.

Swaminathan clarified in several of his above-mentioned publications that Green Revolution was triggered by the genetic manipulation of yield in wheat, rice, maize, etc., whereas the Evergreen Revolution would be triggered by farming systems that can help produce more, from the available land, water, biodiversity and labour resources, 'without' causing ecological and social harm. The emphatic need is to avert ecological degradation so that productivity would remain undiminished over long periods of time.

Swaminathan had outlined the philosophy, principles, objectives related to sustainable agriculture and rural development in his Address titled, *Environmental Protection and Livelihood Security of the Rural Poor*, delivered on the occasion of the Second Indira Priyadarshini Memorial Lecture [146]. Some of the statements relevant to sustainable agriculture and rural livelihoods in his Address are:

Physical availability of food is no longer the principal food security challenge. Providing economic access to balanced diets is the most serious challenge.

Referring to the several steps taken by the Government to enhance the access to food security, Swaminathan cites examples such as Food for work, National Rural Employment and Employment Guarantee to Rural Landless Programmes, the Maharashtra Employment Guarantee Scheme and Mid-day meal programme for school children of the Tamil Nadu Government, etc. He goes on to make the following statements:

In spite of these measures, the degradation of the environment is progressing at a fast pace. Population pressure on land is growing. The holdings are getting smaller and smaller. All the components of the biosphere are under severe stress. Because of inadequate opportunities for gainful employment in rural areas, landless labour families are forced to migrate to towns and cities. Lack of fodder in their native environment is compelling more people with farm animals to take to a 'nomadic' life. Rural women and children have to walk long distances to fetch fuel wood, fodder and water. Flash floods occur whenever there is heavy rainfall. These factors need careful attention, if we are to ensure both nutrition security and livelihood security.

Globally, we have to promote a deep awareness of the ecological interdependence of our planet, as emphasized by the WCED Chaired by Mrs. Gro Harlem Brundtland. We must acknowledge and respect the rights of our environment. Above all, we should break the vicious circle whereby environmental degradation and poverty are inextricably linked as stressed by Indira Gandhi in 1972.

Professionally, I recognize five major areas which need study and understanding if we are to implement effectively the suggestions of the WCED.

For the sake of convenience, I have been referring to them as the five 'Es'. They arise in the spheres of:

- *ecology,*
- *economics,*
- *energy,*
- *employment and,*
- *equity.*

Swaminathan's Evergreen Revolution applauded by Wilson [137] puts ecology in the centre of the development ethos. Social demands and economic goals need to be constrained within the limits of natural resource base. In other words, it will be disastrous to ignore the 'limits to growth' set by finite non-renewable resources as well as those only slowly renewable.

Writing on ecological challenge, Swaminathan refers to the intricate relationships among living organisms and between life and its physical setting. He referred to the great biogeochemical cycles of our planet. This has been further elaborated with scientific data with reference to planet's nitrogen cycle in 2009 in the paper entitled, *A Safe Operating Space for Humanity* by Rockström *et al.* [147]. Too much of chemical synthesis of inorganic nitrogen compound, ammonium nitrate, has led to its accumulation on planet's land and aquifers without a corresponding denitrification process. Normally in nature, the nitrogen-fixing bacteria on one side, and the denitrifying bacteria which convert the excess nitrates back into nitrogen on the other, have been keeping the nitrogen-cycle in balance. With the introduction of Haber–Bosch chemical synthesis of ammonium nitrate, atmospheric nitrogen is more readily converted into nitrates, but there is no corresponding chemical process for denitrification.

Considering the ecological challenges to economic development Swaminathan stated in the Second Indira Priyadarshini Memorial Lecture [146]: *The challenge to ecologists lies in showing how economic improvement can be speeded up within the confines of ecological ground rules. Referring to the economic challenges, he notes that because of the past*

Placing Environment in the Centre of
Economic Development

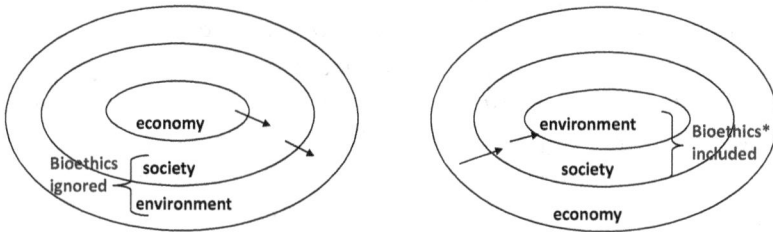

*Consciousness regarding human–nature relationships, social responsibility

human activities, irreplaceable environmental resources have been lost in the course of development processes that, quite frankly got their economics wrong. He went on to say: *we need a new ecological economics that values the living resources of our planet better. So long as the economy is the focal point, bioethical and social dimensions are ignored; it is now agreed that social and ecological dimensions have been weak in the Millennium Development Goals.*

The Sustainable Development Goals (SDGs) which commenced in 2016 are designed to lay greater emphasis on the ecological and social dimensions of sustainable development [148]. It discusses the 'Sustainable Development Goals' (SDGs) in general, and the Goal 2 dealing with *End hunger, achieve food security and improved nutrition, and promote sustainable agriculture,* in particular. It suggests how Evergreen Revolution would help in achieving the Goal 2. It also identifies challenges and threats in an era of climate change and the ways in which science and technology could be harnessed to tackle these problems. There is, however, no magic wand to make SDGs a greater success than the MDGs and rescue the humankind and planet Earth from what now appears to be an imminent catastrophe. During the past five decades, many outstanding scientists and intellectuals have been studying various options to move forward, barring a few who either downplay the environmental issues or even totally ignore them. Focusing on economic growth with utter disregard to the conservation and sustainable use of the natural resources would eventually lead to the *Limits to Growth* and doom. In his famous Coromandal Lecture titled, *Agriculture on Spaceship Earth* Swaminathan [149] cites a passage from article entitled, 'A blueprint for survival' [150] which stated: *The principal defect of the industrial way of life with its ethos of expansion is that it is not sustainable. Its termination within the lifetime of someone born today is inevitable unless it continues to be sustained for a while longer by an entrenched minority at the cost of imposing great suffering on the rest of mankind.*

Swaminathan [149] observes: *In the consumption-oriented world of today, the rich nations consume most of the world's products, while within developing nations the richer sections of society, which may constitute only a fraction of the population, consume a large proportion of the goods produced. It is probable that the garbage cans of the richer nations may contain more protein than the total quantity of protein consumed by the poor of the developing world.*

Swaminathan's explanation for the interesting choice of the title of his Coromandel Lecture [149] (i.e. *Agriculture on Spaceship Earth* in 1972) is an appropriate prelude to Brundtland's report *Our Common Future* [140] that was published 14 years later. He said in the Coromandel Lecture [149]: *The expression 'Spaceship Earth' was first used by Buckminister Fuller many years ago to dramatise the fact that the resources of Earth are finite. Today this viewpoint has become necessary for man. If one looks at it this way, one can see Earth as an extra-large, very efficient, Nature-designed spaceship carrying its full load of three billion passengers (as was in 1973, today in 2016, it is over 7.5 billion) hurtling through space at 18.5 miles per second, equipped with its own energy, refuelling, recycling and life-support systems.*

A thin envelope of air, soil and water around the earth together with energy from the sun provide the major ingredients for the evolution and survival of the large number of living things including man, on earth, and all are closely interrelated. This complex and delicately balanced web of relationships together constitute, the 'biosphere' within which man has his place. It is hence evident that man can ultimately make only a finite claim on the limited resource of Earth.

A highlight of his Coromandel Lecture was his emphatic statement: *We are fortunately in a position to build a positive policy of economic ecology based on a series of 'Dos' rather than 'Don'ts'.* He clarified his statement by elaborating it as follows: *The environmental policy advocated in the richer nations is designed to protect the high standard of*

living resulting from the unprecedented growth in the exploitation of natu-
ral resources during the last century from serious damage by the very
processes of such growth. It is of necessity a policy based on a series of
DONTs. This is inevitable since the aim is to undo some of the damage
already done or to prevent further damage along the same lines.

The poorer nations, however, are faced with the desire and need to
produce more food from hungry soils, more clothing and more housing.
They are aware that historically a rise in standard of living has depended
on the ability of agriculture to release man power to other more industrial
pursuits. They hence naturally wish to develop more industries and to find
productive and remunerative employment for their growing population.
For them, the conditions of poverty and inadequate arrangements for
human and other waste disposal may be greater causes of water and air
pollution than the effluents from factories or fertilizer from the fields.
Since the causes of pollution are by and large different, the solutions will
have to be different too and it would be a grave mistake to attempt to copy
the policies now being propagated in the developed world.

Finally, in my view, Swaminathan through the Coromandel Lecture
[149] was sharing relevant aspects of his thoughts, for shaping the
Evergreen Revolution. The Evergreen Revolution was initiated in the
fields of a few villages in Puducherry, Union Territory (UT) and then in
some of the villages of Tamil Nadu as early as 1988–1989. The villages
where MSSRF started working, initiated a 'systems approach' to sustainable agriculture and rural development from day one. Nothing was, however, forced upon the resource-poor rural women and men in a 'top-down'
manner; instead, it was completely 'bottom-up' and participatory in
nature. The MSSRF scientists strictly adhered to the footsteps of
Swaminathan in interacting with the tribal and rural communities and
learning from them as much as they taught them, or better technologically
empowered them through 'techniracy'. The MSSRF scientists gained a
great deal from the rural communities, particularly the women, who have
tremendous traditional knowledge. Further, their daily lives and ethos are
guided by their ecological prudence. The way that skill and knowledge
empowerment transformed the lives of the rural people on one hand, and
concurrent attention to soil fertility and soil health, management of freshwater resources for agriculture including farm animals, conservation and

enhancement of biodiversity, production and effective utilisation of renewable energy, using the agricultural waste for production of paper, board, etc., instead of incinerating them, and enhancement of economic power and social esteem of the rural women on the other, form the foundation of sustainable rural development. Eco-agriculture (i.e. agricultural practices which are benign or even beneficial to the ecological foundations and public health) and eco-technologies (i.e. technologies for rural eco-enterprises for income-generation) are the two major components of the Evergreen Revolution. These developments on the field were his blueprints of a systems-based sustainable agriculture and rural development. Convinced of its success, he wrote the book [142] *Sustainable Agriculture: Towards an Evergreen Revolution.* The 'paradigm shift' from the Green Revolution to Evergreen Revolution has been explained [151] as below:

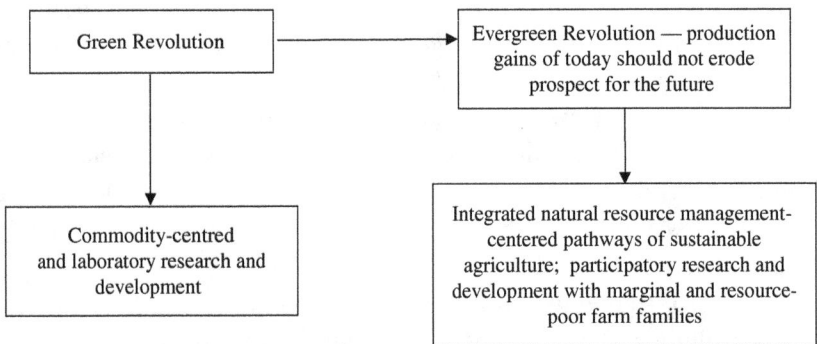

```
┌─────────────────────┐          ┌──────────────────────────────┐
│                     │          │ Evergreen Revolution — production │
│  Green Revolution   │─────────▶│ gains of today should not erode  │
│                     │          │    prospect for the future       │
└─────────────────────┘          └──────────────────────────────┘
          │                                      │
          │                                      │
          ▼                                      ▼
┌─────────────────────┐          ┌──────────────────────────────┐
│  Commodity-centred  │          │ Integrated natural resource management- │
│ and laboratory research and │   │    centered pathways of sustainable    │
│     development      │          │ agriculture; participatory research and │
│                     │          │ development with marginal and resource-│
│                     │          │      poor farm families          │
└─────────────────────┘          └──────────────────────────────┘
```

I joined MSSRF as the Homi Bhabha Chair in Nuclear Sciences and Sustainable Rural Development in 1999. In addition, I was also the first Executive Director of MSSRF. These two positions helped me to visit several villages in Puducherry, Tamil Nadu, Odisha and Kerala and learn the Evergreen Revolution in operation. In close interaction with Swaminathan, a detailed analysis of the structure and functional aspects of the components of Evergreen Revolution was made. A few of the publications entitled *Future Direction of Asian Agriculture: Sustaining Productivity without ecological degradation,* [152] and *Strategies and Models for agricultural sustainability in developing Asian countries* [153] dealt with the components of the Evergreen Revolution and the inter-relationship among the environmental and social as well as gender issues, respectively.

Kesavan and Swaminathan [152, 153] have described the four components of Evergreen Revolution as follows:

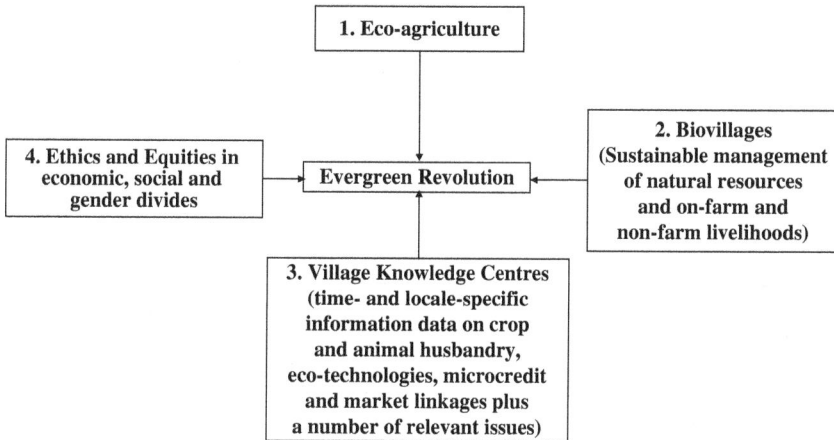

```
                        ┌─────────────────────┐
                        │  1. Eco-agriculture  │
                        └─────────────────────┘
                                   │
                                   ▼
┌──────────────────┐   ┌─────────────────────┐   ┌──────────────────────────┐
│ 4. Ethics and    │   │                     │   │  2. Biovillages          │
│ Equities in      │──▶│ Evergreen Revolution│◀──│ (Sustainable management  │
│ economic, social │   │                     │   │  of natural resources    │
│ and gender       │   └─────────────────────┘   │  and on-farm and         │
│ divides          │              ▲              │  non-farm livelihoods)   │
└──────────────────┘              │              └──────────────────────────┘
                        ┌─────────────────────────────┐
                        │ 3. Village Knowledge Centres │
                        │ (time- and locale-specific   │
                        │ information data on crop     │
                        │ and animal husbandry,        │
                        │ eco-technologies, microcredit│
                        │ and market linkages plus     │
                        │ a number of relevant issues) │
                        └─────────────────────────────┘
```

From the illustration, it is evident that the Evergreen Revolution would ensure food and nutrition security to the millions of subsistence farmers, landless communities and marginal fishers, and necessarily involved concurrent attention to eco-agriculture, sustainable rural livelihoods, knowledge and eco-technological empowerment of the rural cultivators-cum-conservers-cum consumers. And, Ethics and equities across social, economic and gender divides were addressed through the formation of the self-help groups (SHGs). The components of the Evergreen Revolution vastly differentiate it from the Green Revolution. Unlike the Green Revolution that established food security at the national level, but not really at every household level, the Evergreen Revolution aims at ensuring the food security at the individual household level. In other words, Evergreen Revolution is designed in such a way as to fight both the famines of food and rural livelihoods. Since the Evergreen Revolution involves eco-agriculture to achieve productivity in perpetuity, and eco-technologies-driven eco-enterprises to enhance rural livelihoods, these two are discussed in some detail. Further, skill and knowledge empowerment of the rural women and men was done with the help of 'Biovillage' (bios = living) paradigm and the modern ICT-based Village Knowledge Centres (VKCs), respectively. Hence, these are elaborated to an extent:

• Eco-agriculture for Evergreen Revolution

Referring to eco-agriculture, there are several forms of it, from the most 'organic agriculture' to the more flexible, which allows a certain small proportion of application of inorganic chemical fertilisers and 'synthetic chemical pesticides as in the case of integrated nutrition management (INM) and integrated pest management (IPM),' respectively. A few of the major forms of eco-agriculture have been presented in a Guest Editorial by Kesavan and Swaminathan [154]. The same is presented here as a flow-diagram illustrating various terms and their pathways.

Flow-diagram illustrating various terms and pathways of eco-friendly agriculture.

Green Revolution	Evergreen Revolution
Commodity-centred increase in productivity.	Productivity in perpetuity without ecological harm), 'Systems approach'.
↓	↓
(i) Change in plant architecture, and 'harvest index'.	(i) **Organic agriculture:** Cultivation without any use of chemical inputs like mineral fertilisers and chemical pesticides.
(ii) Change in the physiological rhythm-insensitive to photoperiodism.	(ii) **Green agriculture:** Cultivation with the help of integrated pest management, integrated nutrient supply and integrated natural resource management systems.
(iii) Lodging resistance.	(iii) **Eco-agriculture:** Based on conservation of soil, water and biodiversity and the application of traditional knowledge and ecological prudence.
	(iv) **EM agriculture:** System of farming using effective microorganisms (EM).
	(v) **White agriculture:** System of agriculture based on substantial use of microorganisms, particularly fungi.
	(vi) **One-straw revolution:** System of natural farming without ploughing, chemical fertilisers, weeding and chemical pesticides and herbicides.

7.2 Eco-Friendly Cultivation Methods

* *Organic Agriculture*

The International Federation of Organic Agriculture Movements (IFOAMs) has defined organic agriculture [151] thus:

Organic agriculture is a production system that sustains the health of soils, ecosystems and people. It relies on ecological processes, biodiversity and cycles adapted to local conditions, rather than the use of inputs with adverse effects. Organic agriculture combines tradition, innovation and science to benefit the shared environment and promote fair relationships and a good quality of life for all involved.

Four principles govern the identification of organic agriculture. The first is the *principle of health*, which emphasises that the health of all living systems and organisms from the smallest in the soil to human beings are mutually dependent. For example, consuming a diseased fruit or meat of a diseased animal affects the health of the consumer as well.

The second is the *ecological principle* which stipulates that organic agriculture should be based on living ecological systems and cycles, should work with them, and help to sustain them.

The third is *the principle of fairness* directing that organic agriculture should be built upon relationships that ensure fairness, equity, respect, justice in human relations and between humans and other living beings. It insists that animals for milk, meat or draught purposes are provided with conditions and opportunities of life that accord with their physiology, innate behavioural characteristics and well-being. *In fact, the dictum is that organic production systems should be constrained by the animal's needs and not the other way around.* Improvement of quality and quantity of animal products through modern scientific tools and technologies which adversely affect the integrity of the animals is not just acceptable in organic farming. An example of unethical violation of animal welfare has been the use of modern recombinant DNA technology to produce leaner meat in the 'Beltsville pigs'. The genes for human growth hormone were genetically engineered into these pigs to accelerate growth without accumulation of fat in their tissues. These transgenic pigs suffered several health problems such as lameness, ulcers, cardiac diseases and reproductive disorders. Reports of this kind as well as those of allergic reactions in some of the consumers of transgenic canola (rapeseed) are the probable reasons to completely exclude genetically manipulated (GM) crops from the purview of organic farming.

The fourth is the *principle of care* which stipulates that organic agriculture should be managed in a precautionary and responsible manner to protect the health and well-being of the current and future generations of mankind as well as that of the environment. Here, the precautionary approach for decision making recognises that, even when the best scientific knowledge is used, there is often a lack of knowledge with regard to future consequence and to the plurality of values and preferences of those who might be affected. The emphasis is on precaution and responsibility and not on risk assessment which is considered 'a narrow notion based on narrow scientific or economic appraisal'. Thus, it does not permit use of any chemical agents (i.e. fertilisers, pesticides, etc.) or transgenic crops in the schedule of organic farming.

Organic certification is a very rigorous one, and consequently, even a very slight deviation from or compromise with the stipulations in the production of organic foods results in their outright rejection. In many countries, certification is a serious matter of legislation, and commercial use of the word 'organic' outside of the rigorous certification framework is illegal.

- **Green Agriculture**

Green agriculture is quite distinct from Green Revolution. In view of extreme restrictions in practice and uncertainty in achieving high yield potentials within the provisions of organic agriculture, a more flexible and yet ecologically benign system of agriculture had been proposed. This is called 'green agriculture'. Green agriculture is a system of cultivation with the help of IPM, INM and integrated natural resources management (INRM) systems. This is widely practised and promoted in China. The important feature of green agriculture is that it does *not* exclude the use of minimum quantities of essential mineral fertilisers and chemical pesticides. It also permits the inclusion of GM crops. It accepts the view that Bt-transgenic crops would reduce applications of chemical pesticides, and to that extent, it is eco-friendly. However, there are other issues such as biosafety, environmental impact and acquisition of resistance to Bt-biopesticide in the GM Bt crops.

Green agriculture offers several advantages in an era of climate change, especially where nearly 80% of small farms are located in rain-fed regions. It also suits the reclamation of wastelands. The built-in flexibility of green agriculture would help to enhance productivity of small holdings without damage to their long-term production potential. Crop-livestock integration and introduction of stem-nodulating legumes or pulse crops in the rotation system will facilitate the building up of soil fertility. While a large proportion of nitrogen, phosphorus and potash (NPK) could be applied in the form of legumes, farmyard manure, vermicompost, etc., a small, judicious proportion of inorganic fertilisers could also be used to complement the deficit, in requisite levels, for a given crop.

Genetic shielding against abiotic stress (e.g. salinity, floods, drought, etc.) arising from global warming-related sea level rise and climate change is essential for sustainable agriculture. For example, sea level rise is of enormous threat to coastal agriculture due to salinisation of soil and fresh water sources. In order to sustain coastal agriculture with rice as the major staple cereal, genetic shielding of the coastal cereal crops with salinity-tolerant genes from related landraces should be part of anticipatory research. The salinity-tolerant rice strains occur in nature and in using them, conventional plant breeding could lead to developing universally acceptable varieties. At some stage, in the conventional breeding programme, the participation of the local farmers could be involved to develop participatory breeding programme.

- *Eco-agriculture*

McNeely and Scherr [155] have proposed 'eco-agriculture' as a strategy to feed the world while saving wild biodiversity. Eco-agriculture aims at mutually reinforcing relationships between agricultural productivity and conservation of nature. It is defined as an approach that brings together agricultural development and conservation of biodiversity as explicit objectives in the same landscape.

Swaminathan's approach to putting eco-agriculture in practice is inclusive of farm animals, rotation of crops which are cereals, oil seeds, pulses, vegetables, fruits crops and fodder. This wide spectrum of eco-agriculture consisting of farm animals and a variety of crops provide a greater livelihood security to the resource poor small and marginal farmers. The possession of cattle in India is egalitarian and in fact, there are more animals than the fodder and forage produced in the country for them. Consequently, many of the landless women and men owning cows, buffaloes, sheep and goats are nomadic, constantly moving from one place to another in search of fodder for the animals. In such a situation, the resource-poor small farms can gain from keeping animals along with agri-horticultural crops. The urine and dung of the animals contribute to the fertility of the soil. The dung is also used to produce methane that is used as cooking gas in rural households. This also checks the emission of greenhouse gas. Further, the farm animals provide milk, meat and draught power for farm operations. Swaminathan also recommended the inclusion to farm fishes wherever feasible, in the coastal areas. In many of the villages, where MSSRF has linked ecological security with sustainable livelihood security of the resource-poor marginal farmers, the following eco-agricultural practices are going on:

- Fish–livestock–poultry–agro-forestry–vegetables
- Fish–livestock–crops
- Fish–livestock–vegetables–banana
- Aquaculture-based integrated farming system
- System of rice intensification (SRI)–organic recycling–dairy–biogas–vermicompost–forage
- Vegetables–paddy–vermicomposting–IPM–seed production in onion and tomato–SRI
- Low external input-based sustainable agriculture–IPM–organic growth promoters like *Panchakavya*, *amirthakaraisal*, etc.
- Maize–vegetables-integrated pest and disease management–soil health management

There are also few other forms of eco-agriculture. These are as follows:

- *Effective Microorganisms (EM agriculture)*

Dr. Teruo Higa [156] narrates amazing gains in agricultural productivity by application of effective microorganisms (EM). It also solves environmental and health problems associated with chemical fertilisers. EM comes in four varieties which are numbered EM#1 through EM#4. Each type has distinct features and properties. EM#2 features mainly Gram-positive actionomyces; the major content of EM#3 is photosynthetic bacteria, and of EM#4, lactic bacteria and yeasts. EM#1 is a mixture of all the three and exhibits all the three properties found in EM#2, #3 and #4. Each of the four types is more appropriate to certain uses, the appropriateness of each depending on the activities of the dominant species of microorganisms in the mix. The other features of the EM system of agriculture are that rice cultivation involves direct planting without any tilling and weeding and untreated cow dung forms the bulk of the fertilisers. EM agriculture improves the quality of fruits and reduces the cost of external inputs by about one-fifth. EM also turns barren soil into rich, fertile land again.

- *One-Straw Revolution*

Masanobu Fukuoka [157] has described a system of natural farming which has four cardinal principles:

1. No ploughing or turning the soil.
2. No chemical fertiliser or prepared compost.
3. No weeding by tillage or herbicides (weeds play a part in building soil fertility; they need to be controlled, but not eliminated).
4. No dependence on chemical or poisonous pesticides.

With reference to the first point (no cultivation), experience has shown that earth cultivates itself naturally by means of penetration of plant roots and the activity of microorganisms, small animals and earthworms. Left to itself, the soil maintains its fertility naturally in accordance with the orderly cycle of plant and animal life. Weeds are believed to play their part in building soil fertility and in balancing the biological community. This form of eco-agriculture is relevant to carbon sequestration, so essential in an era of climate change.

- *White Agriculture*

White agriculture [151] is a system of agriculture based on substantial use of microorganisms, particularly fungi. The concept of white agriculture took shape in 1986 in China. White agriculture refers to the white-coated scientists and technicians performing high-tech processes to produce food directly from microorganisms or to use them to augment and improve green agriculture.

Transforming the Green Revolution into eco-friendly agriculture for an Evergreen Revolution using one or more of the several pathways described above will usher in a win-win situation for both farmers and ecosystems. Further, it will be prudent to develop for each farm an eco-friendly cultivation plan based on an appropriate combination of different approaches which together can ensure both ecological and economic sustainability.

There is a fundamental difference between the eco-agriculture put into practice by Swaminathan and those of others. For example, the models suggested by McNeely and Scherr, Teruo Higa, Masanobu Fukuoka do not take into account the livelihood dimension of the resource poor small and marginal farmers. Their models to a large extent involve cultivation of crops without destruction of natural biodiversity and also use of non-chemical methods of fertilisation and protection of crops against pests and diseases. There is no inherent mechanism to augment the income generation. In that sense, eco-agriculture is not linked with livelihood security. Swaminathan's approach integrates ecological agriculture with sustainable livelihood security. This approach is therefore highly relevant to India and other developing countries where the food insecurity at the household level is due to lack of purchasing power (access), which is a function of livelihoods/employment. It is for these reasons that, Swaminathan [144] has elegantly put forward the way forward in the following words:

What nations with small farms and resource-poor farmers need is the enhancement of productivity in perpetuity without associated ecological or social harm. The Green Revolution should become an Evergreen Revolution rooted in the principles of ecology, economics, and gender and social equity.

7.3 Eco-Agriculture of Swaminathan: Pathway to Green Economy

The irony today in the Indian agriculture is that many professional agricultural scientists who agree that chemical intensification has done more damage than good in the long term, seem, however, inextricably entrenched in it; number of factors such as mindset, pressure from agro-chemical industries, etc., could be responsible for their retrograde notions. One of their common yet unfounded criticisms against any form of eco-agriculture is that the crop yields would greatly diminish without the use of chemical fertilisers and chemical pesticides and the food production and its availability in the country would substantially fall short of the current and future needs. There are reasons to believe that the current rate of environmental degradation coupled with climate change leading to extreme natural events, would wipe out most of the humans living in the coastal areas as well as in Island states as also the land suitable for cultivation. The Mother Earth has far exceeded her patience and a revenge is imminent. Besides addiction to chemical intensification of agriculture, human population growth rate well beyond the 'carrying capacity' of the planet and unsustainable lifestyle have also been accelerating the degradation of the production base and the production system. *Above all, there is no evidence that eco-agriculture over long periods diminishes productivity; in fact, the contrary is true.* This is discussed later in this chapter.

After Swaminathan [142, 144] developed the Evergreen Revolution and operationalised it in several villages, several other scientists across the world have also come forward with very strong support for sustainable agriculture in one or another form. Many scientists agree that as against the 'exploitative' agricultural practices followed in industrial farming for 'immediate' gains, the Evergreen Revolution incorporates the elements of both 'resilience' (the capacity of systems to buffer against shocks and stresses) and 'persistence' (the capacity of the systems to continue over long periods). It also addresses many wider economic, social and environmental issues. Today, in the new millennium hundreds of well-meaning agricultural scientists throughout the world endorse the Evergreen Revolution of Swaminathan as E.O. Wilson did in his book *The Future of*

Life [137]. All these eco-agricultural scientists recognise the following key principles for sustainability:

- Integrating biological and ecological processes such as nutrient cycling, nitrogen fixation, soil regeneration, allelopathy, competition, predation and parasitism into food production processes;
- Minimising the use of those non-renewable inputs that cause harm to environment or to the health of the farmers (example, families of Punjab farmers suffering from cancer in excess of the spontaneous levels due to excessive contact with chemical pesticides);
- Making productive use of the knowledge and skills of farmers, thus improving their self-reliance and substituting human capital for costly external inputs and;
- Making productive use of people's collective capacities to work together to solve common agricultural and natural resource problems, such as pest, watershed, forest and credit management;
- Community agro-biodiversity conservation and its sustainable use as well as the banks with a difference (Gene–Seed–Grain–Fodder –Water).

Since sustainable agriculture makes the best use of nature's goods and services, Swaminathan [132] advocates eco-technologies that are specifically adapted to local requirements. There are several cases where shifts in factors of agricultural production such as from chemical fertilisers to nitrogen-fixing legumes, from chemical pesticides to natural enemies (Kesavan and Malarvannan [158]) and from ploughing to tillage, have contributed significantly to sustainable agriculture.

Swaminathan's Evergreen Revolution includes, wherever necessary, eco-agriculture involving what is referred to as 'sustainable intensification'. The word 'intensification' in agriculture is not necessarily outrageous for sustainable eco-friendly agriculture. In October 2009, the Royal Society of London published an authoritative and balanced report entitled, *Reaping the benefits: science and sustainable intensification of global agriculture*. It acknowledges that food security is one of this century's key global challenges. This needs to be met in the face of changing consumption patterns, the impacts of climate change, growing scarcity of freshwater and arable land as a diminishing resource. Yet, the crop production methods should sustain the ecological foundations of sustainable agriculture by preservation of resources and support livelihoods of farmers and rural populations around the world, particularly in the developing countries. These various challenges and compulsions thereof *necessitate the sustainable intensification of global agriculture in which yields are increased without adverse environmental impact and without adding more land for farming. Under these circumstances, Swaminathan's Evergreen Revolution that utilises a systems approach including better management of soil and its health, use of biologically-derived than chemically synthesised inputs is the way forward to sustainable intensification. With increasingly diminishing dependence on the chemical inputs, beneficial soil microbes, earthworms, etc., are making significant reappearances.* All these benefits arise from Swaminathan's Evergreen Revolution. *Reliance on fossil fuel-based inputs and farm operations is also substantially reduced in Evergreen Revolution.*

Sustainable intensification also requires several types of resource-conservation technologies and practices that can be used to improve the stocks and use natural capital in and around agro-ecosystems. In his several elaborated writings, Swaminathan refers to *biological softwares* (meaning biofertilisers, biopesticides, etc.) which will be discussed in some detail in the following section of eco-technologies for rural on-farm and non-farm livelihoods.

Broadly speaking, IPM, INM, conservation tillage, conservation agriculture, agro-forestry, rainwater harvesting, storing and its sustainable use, farm animals in resource-poor small and marginal farms are all integral components of the systems approach-based Evergreen Revolution. In the context of eco-agriculture, a distinction between *conservation tillage* and *conservation agriculture* based on the FAO definition is as follows:

Conservation agriculture maintains a permanent or semi-permanent organic cover. This can be growing crop or dead mulch. Its function is to protect the soil physically from sun, rain, wind and to feed soil biota. The soil microorganisms and soil fauna take the tillage function and soil nutrient balancing. Mechanical tillage disturbs this process. Therefore, zero or minimum tillage and direct seeding are important elements of conservation agriculture. Science-based crop rotation further helps to fix

nitrogen in the soil and also avoids disease and pest problems. The three
key principles of conservation agriculture are: (i) permanent residue soil
cover, (ii) minimal soil disturbance and (iii) crop rotation.

Crop rotation also enhances income security to the resource-poor
small and marginal farmers on one hand, and introduction of Swaminathan's
recent concept of 'Farming System for Nutrition' (FSN) [159]. The FSN
seeks to provide agri-horticultural remedies to nutritional maladies in
a given agro-climatic region. The year 2014 had been declared as the
'International Year of Family Farming' (2014 IYFF) by the United Nations
General Assembly at its 66th Session in 2011. The family farming all
over the world received a boost to varying degrees, but its favourable
impact on the Evergreen Revolution is absolutely fantastic [160]. They
have elaborated that smallholder family farms with *pro-nature, pro-poor*
and *pro-women* orientation provide a boost to Evergreen Revolution. They
have shown that revitalisation of family farming traditions with emphasis
on the empowerment of women and youth will enhance small farm productivity and profitability on the one hand, and nutrition-sensitive agriculture on the other. Family farms also enhance rural livelihoods to provide
greater access to food. Corporate farming displaces three to four jobs for
every single job created. Mono-cropping largely practised by the corporate farms is not conducive to develop strategies to provide agricultural
remedies to nutritional maladies in different agro-ecological regions.

An unsubstantiated criticism of eco-agriculture forming the premise
of the smallholder family farms is that productivity increase (kg/ha) in
crops cannot be achieved in smallholder farms. This criticism is untenable and false as shown by Pretty [161] in his paper, *Agricultural
Sustainability: Concepts, Principles and Evidence.* This paper presents
huge amount of fantastic data to support eco-agriculture. He has dealt in
detail with the effects of sustainable agriculture (i.e. eco-agriculture of
Evergreen Revolution) on yields. A very large study comprised the analysis of 286 projects in 57 countries. It was found the mean relative yield
increase was 79% across a wide variety of systems and crop types.
Further, the smallholder farms practice sustainable intensification with
'green' inputs (biofertilisers, biopesticides) and largely renewable energy
(cow dung, solar and wind) in contrast to corporate farms which use

fossil fuel energy based inorganic chemical fertilisers, diesel-driven trac-
tors, driers, pumps, harvesters, etc. The process of achieving the yield
gains through the sustainable intensification is slower than in the case of
chemical-based intensification of agriculture, but are quite long-lasting
and sustainable.

Further, the large chemically intensified corporate farms have little
space for women farmers. On the other hand, the family farms have tradi-
tionally placed a heavy share of workload and responsibility on the shoul-
ders of women farmers. Yet, in India, these women do not have the right
to land. The women-managed small family farms (i.e. these could as well
be subsistence farms) are handicapped for credit, technological support,
etc. When the women-operated/managed family farms fail, the result is
the 'feminisation' of poverty and undesirable social consequences follow.
Hence Swaminathan who was a Member of the 'Rajya Sabha' introduced
a Bill, *The Women Farmers' Entitlements Bill*, 2011 in the Rajya Sabha on
May 11, 2012. Its purpose is to provide for the gender-specific needs of
women farmers, to protect their legitimate needs and entitlements, and to
empower them with rights over agricultural land, water resources and
other related rights and for other functions relating thereto and for matters
connected therewith. There can be no families without women and no
family farms either.

One of the major disadvantages of the smallholder farms is the lack of
power of economy of scale. On the criterion that smallholder farms largely
practice 'near organic' or 'green agriculture' following the INM and IPM,
it is expected that their produce should fetch higher prices than those produced in the factory farms. However, that is not the case. Among the many
reasons, one is that the produce is not directly sold to the consumers, but
to traders, who buy from the small farmers at very low prices, and sell
these with several-fold profit to consumers. This problem could be greatly
mitigated through the formation of cooperatives which pool the produce
from individual family farms and manage the marketing aspects. The farmers should get fair prices. The milk cooperative set up by late Verghese
Kurien is an example. Milk is purchased from thousands of rural women
who have a few cows/buffaloes. They take the milk to the collection
centres where they are also immediately paid the cost of milk. It results
in pooling of milk on the one hand, and ensuring livelihood security for

women, on the other. The National Commission on Farmers in its report submitted in 2007 to the Ministry of Agriculture, Government of India, had recommended the promotion of commodity-based farmers' organisations like Small Cotton Farmers' Estates, Small Farmers' Horticulture Estates, Small Farmers' Medicinal Plant Estates, etc., in order to combine the advantages of decentralised production and centralised services, post-harvest management, value addition and marketing for leveraging institutional support and facilitating direct farmer–consumer linkage. Distress sale by small and marginal farmers at discounted prices for immediate cash is a serious problem which could be countered to some extent by liberalising pledge loans against produce stored in godowns/farmer's own home.

The industrial farms growing potatoes and onions could have their own cold storage facilities and they can sell the produce at higher profits when demand is more and supply is less. The smallholder farms cannot afford to do so. They do not even have proper storage facilities. Therefore, the government support to smallholder farms should be to provide cold storage facilities, improving the warehousing systems, facilitating loans against warehouse receipts, etc.

The advantage of cooperatives of smallholder family farms is that it can also open up avenues for undertaking ecological farming techniques like IPM and INRM, group credit, group insurance, improved post-harvest technology, water harvesting and water sharing. In the ultimate analysis, farmers (especially youth from family farms) will need opportunities for assured and remunerative marketing. That alone can sustain farmers' interest in farming, particularly family farming. The farmers' cooperatives, if designed with pro-small farmer and pro-women orientation, will help trigger a nationwide movement. The World Trade Organisation should appreciate that smallholder family farms are important for the following reasons:

- These are inextricably linked to world food security.
- These preserve traditional food products while contributing to a balanced diet and safeguarding the world's agro-biodiversity and the sustainable use of natural resources.
- These also contribute to local economies, and rural livelihoods more substantially than industrial/ corporate farms.

Chapter 8

Launching Evergreen Revolution to Link Ecological and Sustainable Livelihood Security

The previous chapter brings out how Swaminathan's 'Evergreen Revolution' integrates ecology, economics and equities (both social and gender) to render it inherently sustainable. It also addresses the pathways to transform the Green Revolution into an Evergreen Revolution. The paradigm shift from commodity-centric thrust to productivity through a systems-based approach involves concurrent attention to locally adapted crop varieties, soil health, quality of irrigation water and rainwater harvesting, conservation of biodiversity, renewable energy as well as mitigation of greenhouse gases. Swaminathan's conviction right from the very beginning was that agricultural productivity in perpetuity alone without concurrent attention to enhancing the resource base and skills of the rural women and men to create and sustain income-generating livelihoods would not reduce poverty and hunger. Swaminathan's ingenuity in basic scientific research has been already vividly brought out in the earlier chapters; he has yet again emerged absolutely ingenious in creating on-farm and non-farm livelihoods to generate income for the resource-poor marginal farmers, fishing communities and landless rural families. To the best of my knowledge, it was just Swaminathan who had deep concern about the 'access' dimension to food security at the household levels and hence focused on rural livelihoods and developed strategies to provide skill and knowledge empowerment of the largely illiterate and unskilled resource-poor rural women and men. Skill and knowledge are prerequisites for the sustainable livelihoods. Swaminathan was emphatic that lack of literacy should not be an impediment to acquiring skills for undertaking rural livelihoods. He had also reasonably thought of agro-climatic zone-specific on-farm and non-farm livelihoods, besides augmentation of income to the landless rural women from producing *biological softwares* (i.e. biofertilisers, biopesticides, etc.) which also form the green inputs for the eco-agriculture of the Evergreen Revolution. Swaminathan won the highly deserved First World Food Prize in 1987. The prize money formed the initial seed money to set up the M.S. Swaminathan Research Foundation (MSSRF) in Chennai by 1988. He won several other awards too, which carried a financial component. All these monies were utilised to put up his 'magnificent temple of sustainable agriculture and rural livelihoods to achieve Zero Hunger goal'. The monies earned from several awards (the World Food Prize, Tyler Prize for Environmental Achievement, Honda Prize for developing eco-technologies, UNEP–Sasakawa Environment Prize, Volvo Environment Prize, Blue Planet Prize, Indira Gandhi Prize for Peace, Disarmament and Development and several scores of such Awards) form even today the backdrop of the eco-friendly MSSRF buildings on the Third Cross Street of Taramani, Chennai, which house the science of sustainability based programmes. I believe that both the Hindu Goddesses of Knowledge and Wisdom, 'Saraswathi' and the Goddess of Wealth, 'Lakshmi' have together blessed Swaminathan. Such a happening is indeed very rare, as it is said that their blessings are mutually exclusive.

Money is essential and it should be earned by fair means. Equally important, however, is that after meeting the needs of the family and keeping adequate savings, the surplus should be diverted/donated to such purposes as welfare of the less fortunate fellow beings and for rescuing the planet Earth from the brink of annihilation. This is exactly the rationale for Swaminathan, his wife Mina, who is an eminent authority in the field of education with special reference to the preschool education, integrated child development, gender and sustainable development and their three daughters Soumya, Madhura and Nitya each with nationally and globally distinguished careers in medicine, economics and gender science, respectively for donating all the prize monies to the establishment of this foundation. They have also donated their valuable piece of land in Kalpetta, Wayanad, Kerala for setting up MSSRF's site for the 'Community Agro-biodiversity Centre' which is worth quite a few crores of rupees. In MSSRF main building, the auditorium, named after Swaminathan's father Sambasivan, is quite unique. It has something of a reverential inspiration about it. At least, this is my feeling. Just outside of the Sambasivan auditorium, there is fixed a plaque with the following inscription:

> *M.S. Swaminathan Research Foundation Centre for Research on Sustainable Agriculture and Rural Development, April 14, 1993: Dedicated to the use of science for fostering sustainable human livelihood and conservation of nature.*

It is appropriate to recollect the global scenario when the MSSRF was set up in 1988 and then judge its performance record about 20 years later in 2009 by two internationally acclaimed independent reviewers.

While writing MSSRF's history of about 10 years (1990–2000) under the title *Social Vision for Science,* Swaminathan wrote, *MSSRF was established at a time when humankind started facing serious ecological and social problems — growing damage to the basic life support systems of land, water, forests, biodiversity and atmosphere, increasing poverty as well as social and gender inequity; rapid growth in human population resulting in reduced per capita availability of land and water; and explosive technological development coupled with high rates of unemployment, resulting in jobless economic growth.* He went on: *All these factors led to MSSRF defining its research agenda in terms of sustainable development, rooted in the principles of ecology, social and gender equity, employment generation and economic viability. The fostering of a pro-nature, pro-poor, pro-women and pro-employment orientation to technology development and dissemination in rural areas became MSSRF's mission. If technology was an important factor in the past in increasing economic and social disparities and causing ecological harm, MSSRF's approach has been to enlist appropriate blends of traditional and frontier technologies as allies in the movement for economic and ecological well being and gender equity. In the field of agriculture, MSSRF's goal became one of spreading an evergreen revolution based on sustainable advances in biological productivity. This emphasized the need to place efforts to increase crop productivity on the foundation of integrated natural resource management.*

As the Chairman of the Board of Trustees of MSSRF, it was Swaminathan's decision to get an independent review of the programmes and their progress at the end of 20 years of its existence. The two independent reviewers, Drs. Uma Lele and Kavita Gandhi took adequate time to visit MSSRF and its various field sites, talk to the various technical, scientific, administrative and supportive staff, critically review its publications as well as its roles in evolving public policies for the Government of India and made the following points in their report [162].

- *MSSRF has no parallel in its entirety.*
- *It is unique since its research and development endeavours are aimed at finding solutions for the already existing, as well as nearly emerging problems, often resorting to both trans- and inter-disciplinary approaches. That in fact is the core strength for its recognition worldwide. Yet another major cause of its success is its ability to convert laboratory-based scientific findings into field-level applications. Towards this activity, the MSSRF has been actively promoting participatory research with farming families on one hand, and policy research designed to achieve synergy between grass-roots experience and public policy, on the other.*
- Drs. Lele and Gandhi have made a fair assessment of MSSRF's journey of two decades on its path of social vision of science. As to the *Future we want: Journey of MSSRF*, Swaminathan, the Founder-Chairman wrote the following in the 22nd Annual Report, (2011–2012): *MSSRF is designed as an equal opportunity centre for all socially committed scientists, regardless of gender, age, religion, caste or community-verily a 'Vasudhaiva Kutumbakam' in action; it is also a 'centre without walls', generating synergy and symbiosis in partnership with similar institutions with similar goals. Mahatma Gandhi's advice that we should keep our windows and doors open so that fresh ideas come from all directions, but we should keep our feet firmly on the ground, has guided the research philosophy. Above all, participatory research with tribal and rural women and men and anticipatory*

research to scientifically checkmate the adverse consequences of climate change and sea level rise were chosen as pathways for shaping the future we wanted.

- MSSRF: Temple of Learning — Many distinguished visitors to MSSRF are awe-inspired not just by the quality and standard of science fulfilling social contract, but also scores of different approaches, both disciplinary and inter-disciplinary being used to reach the goal. In the visitors' book some have referred to MSSRF as temple of learning. Yet another question is to rate MSSRF among Swaminathan's numerous other creations. It is really difficult, yet MSSRF is the first among the equals; shall we say it is his 'magnum opus'.

Most people agree that Swaminathan and his unique creation MSSRF have been fulfilling the social contract of science and technology. The accelerated economic growth that nations crave for has no elements/ solutions to fight the Malthusian scourge. The policy makers and a few population experts are contended with the possibility of stabilisation of India's and world's human population by 2050, but it needs to factor in the planet's non-renewable resources. The human population of the world before 1 BCE was less than 10 million. The world population reached about 500 million in the year 1650, and it rose to about 950 million in 1798 when Thomas Robert Malthus noted that geometric population growth would inevitably outstrip food production, leaving society destitute and hungry. It is unfortunate that Malthus' *Essay on the Principle of Population* has been long regarded as 'Malthusian Curse' rather than a rational analysis for understanding and family planning. Coupled with either religious or socio-economic as well as cultural excuses, family planning has been consistently neglected. The governments since the independence of India have consistently avoided focus on the family planning; instead, 'vote banks' were encouraged to proliferate. Equally unfortunate was the unfair treatment given to the book *Limits to Growth* published in 1972 by Daniella Meadows *et al.* [163]. About 100 years after Malthus, in 1900, the global population reached about 1.6 billion. In 1967–1968, the year of India's Green Revolution, its population was about 400 million and that of the world about 3.3 billion. Then, India's proportion to world population was about 13% and in 2015, India's population reached about 1,260 million and that of the world around 7,300 million which makes India's share 17.3%. Swaminathan's idea of the Green Revolution was to provide a 'breathing space' to enhance productivity of crops and not for 'baby booms'. The free but unfair trade under globalisation had induced a severe competition that necessarily set aside ecological and social concerns. By the end of the 20th century, bulk of the humankind seemed to have forgotten humanity; of course, ecology and environment had seldom figured in the agenda of several developing countries which had become embroiled in internal conflicts and strife. The famous British astronomer Martin Rees [15] who has visited MSSRF wrote in his famous book, *Our Final Hour, In the absence of substantial reduction in resource consumption and environmental degradation, the odds are no better than 50:50 that our present civilization of Earth will survive to the end of the present century.* The implicit reference to MSSRF was that under the guidance of Swaminathan, it was doing all the best it could to extend the lease on human survival and its civilisation in a beleaguered world. The 'ecological footprint' of the globe has already been exceeded by about 30–35%. With still burgeoning human population and its unsatiable extravagant lifestyle, there is no likelihood to reduce the degree of overshoot. India, with about 17.3% of human population and 15% of the global livestock with only 2.3% of geographical area, 4.2% of freshwater sources, 1.0% of forest area and 0.5% of pasture/grassing areas, is still able to produce adequate food for its people with a surplus. As of now, there is no desperate need to have to resort to unsustainable modes of agriculture. There is, however, a desperate need to fight the famine of rural livelihoods. Further, the most worrisome issue from sustainable development point of view is the environmental degradation, particularly the loss of biodiversity. These twin growing problems receive major attention in MSSRF. In this context, Swaminathan's Leelavathi Memorial Lecture [138] laid the 'Foundation stone' of MSSRF that actually came to be set up after 26 years in 1988–1989.

Professor Jeffrey Sachs of Earth Institute, New York in his foreword to the book *From Green to Evergreen Revolution: Indian Agriculture: Performance and Challenges* [9] wrote, *Few people in modern history have done more to help humanity surmount the Malthusian challenge than Swaminathan.* This is so because, Professor Sachs continues, *even with all our technological wizardry, we have not yet conquered the Malthusian Challenge since we have not adopted a truly sustainable method of feeding the planet* and goes on, *the great agronomic successes since Malthus' time, including the Green Revolution itself, have come at a huge and sometime irreversible environmental costs.* Swaminathan's idea of transforming the Green Revolution into an Evergreen Revolution in MSSRF was precisely to avoid to the maximum extent, wherever feasible, the unsustainable methods, tools and techniques. The ecological and social components of sustainable development receive equal and concurrent attention as the economic component. There is, of course, a small component of recombinant DNA technology-based anticipatory research programme, admittedly not very eco-friendly in approach to sustainable agriculture. The products of this anticipatory research could, however, be useful to support continued cultivation of paddy even after parts of the coastal land and aquifers have become salinised due to sea level rise on account of global warming. In as much as it would enable the paddy cultivation go on in the resource-poor small and marginal farms in an era of sea level rise, it is 'sustainable'. MSSRF also accords priority to study the naturally occurring salt-tolerant landraces.

MSSRF, the 'magnum opus' of Swaminathan is possibly the most unique and singular non-governmental research establishment to set its mission of research and sustainable development in tune with nature. In the book *Thirty years after the 1972 Limits to Growth*, Meadows *et al.* [164] point out the anthropogenic climate change fast approaching the 'tipping point'. In an era of climate change, the acquisition of resistance to pests in the transgenic crops (Bt and Ht) growing areas is a major threat to sustainable agriculture. Genetic resistance development by pests through the enhancement of the 'selection pressure' as with growing the pesticide-producing transgenic (Bt and Ht) crops negates sustainable management of pests. The natural events, viz., mutations and natural selection confer genetic shielding against the chemical pesticides to the ephemeral-generationed pests, whereas the residue of the chemical pesticides is carcinogenic to humans. Therefore, as the Chairman of the Task Force on Agricultural Biotechnology, Swaminathan had advised to accord very low priority to Bt and Ht transgenic crops. With ecological prudence in view, MSSRF's eco-agriculture adopts crop rotation, biopesticides and provision of ideal niche for the predators and parasites to thrive. Swaminathan's wisdom and his wavelength of communication with nature (also possibly to some extent with her secret) indeed constitute the intellectual and scientific resource to the foundation of MSSRF. Therefore, it is my belief that MSSRF cannot be replicated, and its present and future programmes cannot far too greatly deviate from the goals, ideals and approaches envisioned by Swaminathan. It took him decades to refine his innovative ideas and to reach a level of *'sustainable do ecology'* for reconciling the twin goals of conservation and development. It now has a perfect but delicate equilibrium which needs to be maintained and far greatly improved.

MSSRF is not a welfare organisation; it has no mandate and resources to run programmes like the 'Mid-day lunch' at schools, etc., instead, it stands for harnessing science and appropriate technologies for sustainable management of resources in the rural areas, provide skill and knowledge empowerment of the rural women and men to create on-farm and non-farm livelihoods and fruitfully fight the famine of rural livelihoods. One of the challenges is to demystify the laboratory-based high level science-based technologies into absolutely simple and 'doable' ones, not in the sophisticated, air-conditioned laboratories but in 'huts' and small sheds that are clean but not necessarily sterile. Several eco-enterprises include productions of oyster mushroom (*Pleurotus ostreatus*), paper and board from agricultural waste, egg parasitoid *Trichogramma chilonis*, fungicide *Trichoderma viride*, biofertilisers like azolla, vermicompost, etc. Actually, achieving this in practice entails a series of well-thought-out steps involving 'bottom-up' (i.e. participatory) approaches to provide skill and knowledge empowerment to the largely illiterate and unskilled rural women and men. A jewel to Swaminathan's intellectual throne is his humility coupled with his ability to listen to everyone, no matter whether the person is a politician, top official, a scientist with tall claims or just a humble rustic from an extremely backward hamlet.

At this point, I would like to introduce to the readers one of the recent outstanding book, *M.S. Swaminathan in Conversation with Nitya Rao: From Reflections on My Life to the Ethics and Politics of Science* [165]. Professor Nitya Rao, the youngest of the three daughters of Swaminathan and Mina, also has the distinction of being the youngest professor in Gender and Development at the School of International Development,

University of East Anglia, U.K. With humility in heart, I would like to acknowledge the fact that no biography could be truly as revealing and reflective as the autobiography. Nitya, with her remarkably sharp intellect and ocean of experience as a social scientist, has at several places provoked her father, Swaminathan, to speak out his mind and also reflect on several major decisions and actions he had taken in his erstwhile professional career. He has obliged with details of certain things that he had to do at a point of time and situations that led him to think multi dimensionally, temporally, spatially and across different domains of life. At places, Nitya specifically asks him on his ability to connect easily with the past, present and future, recalling the lessons learnt in the past to confront the future challenges to find the links from the local to the global, from practice to policy from people, in particular women and the poor, to environmental, climate and scientific concerns, and from production to distribution to consumption. Nitya has asked a specific question on developing the mandate of MSSRF that is multi dimensional towards the basic goal of linking ecological (i.e. biodiversity from microbes to all other organisms including humans), social (equities), gender (empowerment of women) issues with the economic dimension, namely the sustainable rural livelihoods. Swaminathan's response to Nitya is given below in the form of a brief narrative.

- Environment friendliness, particularly conservation of biodiversity as well as the precious resources such as soil, freshwater, social and gender equities and economic viability, constitute the very basic dimensions of sustainable agriculture. Loss of biodiversity, loss of good soil, over-exploitation of ground water and the anthropogenic climate change have emerged as major challenges/threats to sustainable development enshrined in *Our common agriculture* by Swaminathan Panel referred to in *Our Common Future* [140].
- Swaminathan narrates his personal assessment over the years with reference to the extensive loss of biodiversity, particularly coastal mangroves being cut down for aquaculture ponds.
- Swaminathan chose 'pro-nature' as the first point because natural resources which are finite or only slowly renewable need to be conserved and sustainably managed for achieving sustainable development and reducing the progressively increasing inter-generational inequity.

At this stage, Swaminathan points out that the United Nations would launch Sustainable Development Goals (SDGs) from the beginning of 2016 in place of the Millennium Development Goals (MDGs) which would end by the December 31, 2015. Swaminathan is for this change as the environmental degradation and genetic erosion needs to be strongly arrested than it ever was.

- Inequality today is increasing and in turn, it leads to social violence. Swaminathan cites the Roman philosopher Seneca, (4th century BC–65AD) and emphasises, *A hungry person listens not to reason, nor cares for justice, nor is bent by any prayers. He wants food today.* Keeping this at the back of his mind, Swaminathan identified gender equity and poverty as the most crucial concerns needing immediate and strong action. He was quite touched by the sight of landless labourers working hard without adequate caloric intake, and even worse, those who did not get work were starving. This explains the introduction of 'pro-poor' orientation in all the development endeavours of MSSRF. Yet, the most poignant reality was the social discrimination and violence against women. Swaminathan refers to the inordinate insensitivity, particularly in the rural areas to the problems faced by women and also their special needs. Women in the rural areas walk several kilometres a day to find water and fetch it in pots placed over their heads and hips. Men hardly realise what it takes to get water to cook, to wash, and also provide for farm animals. In this regard, I recollect Mina often pointing out in meetings and conferences about the false notion held by a large section of men that their wives were sitting 'idle' at home. Mina has been right in saying that given the invisibility of women's work, both their contributions to agriculture and domestic responsibilities were seen as 'natural' rather than 'skilled', men do not treat women as their equals, nor do they acknowledge that women play a more vital role in the welfare of the entire family. These aspects of rural life in India made Swaminathan orient MSSRF's research and development programmes and field-based activities in favour of women, explaining its pro-women orientation.

There are two other statements by Swaminathan which throw a great deal of light on the goal of MSSRF and one of its key programmes, namely, the 'Biovillage paradigm'. Swaminathan made it clear: *MSSRF was not designed to be a centre for social transformation per se; rather the goal was to use science and technology to trigger social change. However, we had to be sensitive to the needs of different groups of people.* He also clarified the concept of biovillage and identified the criteria that would transform a typical Indian village into a true biovillage. He explained the three criteria he had in his mind: (i) conservation and enrichment of the natural resources, particularly common property resources, (ii) improving productivity and profitability of the small and marginal farms, and (iii) generating non-farm employment and new kinds of jobs, especially for the youth.

In the initial stages, Swaminathan and a few of his senior colleagues of MSSRF sat with the landless women and men and decided on activities like mushroom cultivation, vermiculture, for which they had no need for land, but which could bring in income. Later, the landless women in several villages of Puducherry, Tamil Nadu, Andhra Pradesh and Vidarbha region in Maharashtra were given the necessary training through *techniracy* to produce in their huts the egg parasitoid *T. chilonis*, a tiny wasp that lays its eggs into the eggs of serious crop pest, mainly borers of fruits, stems and bolls of cotton such as *Helicoverpa armigera* and destroy them. The developing larvae of the wasp feed on the yolk in the eggs of the pest for their own development, killing the pest larvae in the process. The *Trichogramma* egg parasite is reared by landless women; after demystifying the technology, MSSRF has provided the rural landless women with skill, knowledge and initial resources to adopt this as a means of livelihood. Several women have become experts in producing the *Trichogramma* biopesticides. The women collect the eggs of the parasitic wasp on cards which are then stapled to the leaves of the crop plants for protecting them against the borer pests. When the borers lay the eggs on the fruits, stems and bolls, the adult parasites emerging from the eggs stuck on the cards, attack the eggs of the borers and parasitise them. Consequently, the 'economic injury level' is well within acceptable levels. These landless women make money by selling the cards on which the eggs of the parasite are stuck. The economic condition of the landless women often caught in a 'debt trap' substantially improves. The *Trichogramma* production-based livelihood is a good example of Swaminathan's *pro-nature, pro-poor, pro-women* and *pro-rural* livelihood policy for technology development and dissemination to the rural areas. It is so convincing because of the following:

(a) The egg parasitoid biopesticide is rather specific to the target pest, and hence, it does not leave any carcinogenic residue on the edible plant parts on soil and water; so, it is indeed *pro-nature*.
(b) It results in the generation of income to the landless poor; due to the lack of purchasing power, they have remained food-insecure and hungry. Since this technology fetches income to the landless labourers and enables them to access food, it is *pro-poor* in orientation.
(c) It is quite well-known that in the poverty-stricken food-insecure house-holds, it is often the women who face hunger more frequently and also intensely; the hunger and deprivation of women become extremely cruel when their husbands happen to be drunkards. Without money in their hands, and nowhere to go, the women particularly in the rural areas suffer the worst. Hence, Swaminathan's goal, also that of MSSRF, to provide skill empowerment to women and help them to manage an income-generating rural livelihood has significant social and gender relevance. The egg parasitoid technology largely managed by the land-less rural women has thus the *pro-women* orientation.

Over the years, I have been quite impressed with a lot of intangible benefits noted to accrue from the biovillage paradigm. First of all, a biovillage becomes a completely food-secure village. Different eco-enterprises within a village help the village communities consisting of farming, fishing and landless families to become more self-sufficient and self-reliant. These build their self-confidence too. Second, women who work and earn become economically self-reliant; it enhances their social esteem and independence. Further, the economically, ecologically and socially vibrant biovillages with sustainable agri-business enterprises can keep the youth in farming and keep the land grabbers out. The government should spread the biovillage paradigm as it is also an effective antidote to despondence. His ingenuity and deep concern for the poor and hungry combine to drive Swaminathan constantly to design unique methods, tools and techniques ideally suited for the sustainable rural development. For instance, the technologies which work well in the urban factories and the multinational production units are scrupulously avoided for the rural areas. Many of these technologies are indeed at the frontier areas, and in fact, these represent the state of the art. Yet, the fact remains that most of these technologies are two-edged, capable of doing both good and harm. Swaminathan is one who would *not* want any technology to ever cause harm to the rural people; he would want technologies to be always beneficial to the rural people. So, he blended the frontier technologies (be it nuclear, biotech, space, nano or any other) with the traditional knowledge and ecological prudence of the indigenous rural and tribal people, mainly the women. Consequently, the resultant technologies become what are referred to as *eco-technologies* having acquired *pro-nature, pro-poor, pro-women* and *pro-livelihood* orientation. Replacement of chemically fixed nitrates with biologically fixed nitrates in the soil with biofertiliser (e.g. *Azolla* cultured by rural women) or the chemical insecticide with egg parasitoid *Trichogramma* forms good examples of the eco-technologies. While the various steps and processes involved in the production of a biopesticide/biofertiliser or oyster mushroom on paddy straw form the eco-technology, the resultant products represent the eco-enterprises. These constitute on-farm or non-farm-based activities. The eco-enterprises require resources like any other technology-driven enterprises. Therefore, Swaminathan's biovillage paradigm has a major component of 'Natural Resources Conservation and Enhancement'. The biovillage also has a bio-centre for providing skill and knowledge empowerment and helping the resource-poor rural women and men with backward (i.e. loans, etc.) and forward linkages (i.e. market for their produce, etc.). The following two flow diagrams taken from the book *Evergreen Revolution in Agriculture: Pathway to a Green Economy* [151] show the structure of the biovillage and the various facets of the biocentre (Figures 8.1 and 8.2).

It may be recalled that the first UN Conference on Conservation and Development was held in 1992 at Rio de Janeiro. At the Rio Conference, several international conventions like those relating to Biodiversity, Climate Change as well as Agenda 21 with specific guidelines for sustainable development were adopted. Later, a Convention on Desertification was taken up. Twenty years later, in June 2012, the RIO+20 with the focal theme on Sustainable Development and Green Economy was held in Johannesburg, South Africa. In preparation for the RIO+20, Swaminathan made two very major contributions: (i) MSSRF organised an inter-disciplinary dialogue on the theme *Environment, the New Economy and New Employment* at

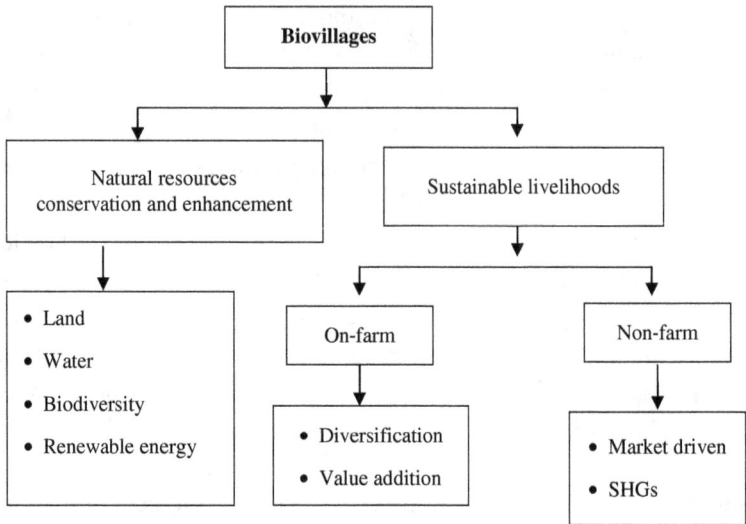

Figure 8.1.   Structure of the biovillage.

Source: Taken from Ref. [151].

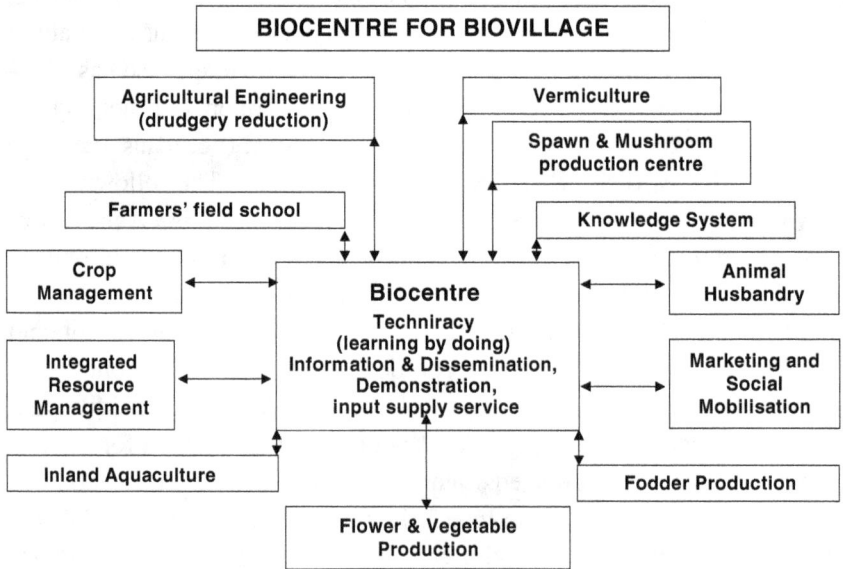

Figure 8.2.   Biocentre for biovillage.

Source: Taken from Ref. [151].

Chennai from January 28–31, 2002 with support from Global Environment Facility. It led to identifying various opportunities for fostering environmentally and socially sustainable job-led economic growth. At that time, many countries including Johannesburg were witnessing youth unrest due to lack of jobs. Swaminathan's intervention by way of introducing new employment opportunities especially for youth was therefore timely. (ii) Swaminathan wrote the book *From Rio de Janeiro to Johannesburg: Action Today and Not Just Promises for Tomorrow* [132] which was highly relevant for the conference. The last para of his Introduction to the book effectively articulates the food and water security as the index of progress towards achieving sustainable development in an era of climate change. The last para reads:

Food and drinking water occupy the first position in the hierarchical needs of human beings. I have therefore, chosen sustainable food and water security as the index of our progress towards achieving the humanistic goals of Stockholm, Rio de Janeiro, and Johannesburg. It is noted that food security includes the sustainable livelihood security. Swaminathan concludes the Introduction with an appropriate verse from George Herbert. It states: *sweet peace cannot be searched, for it grows with grains with which bread is made to quench hunger.*

At this juncture, it is necessary to point out that the biovillage paradigm, one of the flagship programmes of MSSRF, had infused green economy into its activities right from its very beginning in 1995. About 17 years later, the UN Conference 'RIO+20' held in 2002 in Johannesburg, South Africa adopted the goal of green economy, defined as *one that results in improved human well-being and social equity, while significantly reducing environmental risks and ecological scarcities.* This indeed has been the very basic goal and dictum of MSSRF which started its activities towards sustainable development a little over two decades prior to the RIO+20 Conference. This shows that Swaminathan had foreseen things to come and how his plans to tackle the forthcoming challenges and threats are far from the anthropogenic designs but closer to nature's path.

Some of us, including myself, become over enthusiastic about the newer technological capabilities, but there is hardly any technology that has not also turned out to be 'hatchet' in the long run. The most recent example is the technology for harnessing solar energy that is obviously clean and renewable; however, it is now being realised that solar panel waste would emerge as a serious environmental problem; the same has been true for nano-waste, electronic waste, nuclear waste, etc. In all the cases of human-made technologies, the waste management ultimately becomes a real serious problem. On the other hand, the processes which occur in nature are elegantly recycled, leaving no waste at all. Since the era of industrial revolution began with the invention of steam engine in 1780 by James Watt, there has been remarkable progress in technologies in every sphere of human lives — agriculture, medicine, textiles, aviation, automobile and even in arts and music. But they all exert a burden on the recycling processes in nature; these in turn adversely affect human health and environment. The nitrogen cycle that had been perfectly operating in nature without causing accumulation of nitrates in the soil and aquifers has now been irreparably vitiated by the Haber–Bosch chemical process to synthesise ammonia and ammonium nitrate for agriculture and industry. While these are quite invasive, the eco-technologies are far greatly milder and closer to the processes occurring in nature.

In the 1990s, no one had ever thought of eco-technologies and their role in eco-agriculture and *on-farm* and *non-farm* rural enterprises to provide income generation opportunities to the landless rural families, the combination of eco-agriculture in one form or another with various on-farm and non-farm eco-enterprises. However, in trade under globalisation, ruthless competition often involving unfair clauses and practices, the system of 'production by masses' is handicapped and 'mass production' is favoured. On the other side, Professor Wakernaagel and his followers as well as the Asahi Foundation in Japan are deeply concerned over the ever-increasing 'ecological overshoot' [166]. Is not economic growth, like cancerous growth, leading the planet and humanity to its doom? The Earth's capacity to produce our needs of food, fibre and shelter and at the same time absorb and recycle the waste generated has already been far exceeding the growing demand.

Chapter 9

Pioneering the Use of Modern ICT for Knowledge Empowerment

In Chapter 8, reference had been made to the biovillage paradigm with eco-technologies for conservation of resources and creation of sustainable on-farm and non-farm livelihoods in the rural areas. Many of the eco-technologies were already in practice in a few villages when Swaminathan convened a dialogue *Eco-technology and Rural Development* at MSSRF on April 11–15, 1993. Several participants presented papers on the technological and social dimensions and these have been compiled and edited by Swaminathan in the form of a book [167]. The concept of eco-technology was originally developed by late Commandant Jaques Cousteau (the famous under-sea explorer); its ardent promoter Professor Fredrico Mayor, then of UNESCO, greatly admiring the excellent contributions to eco-technology already made by Swaminathan at MSSRF, appointed him to the globally renowned and highly prestigious 'UNESCO–Cousteau Ecotechnie Chair' in 1994. Swaminathan is presently continuing in this position making globally noteworthy contributions to alleviate rural poverty and hunger of the most deprived communities living in the rural areas.

The participants of the Dialogue on *Eco-technology* in 1993 had noted that Swaminathan also had organised an International Dialogue in 1992 on *Information Technology* [168] at MSSRF. It helped standardise methods of taking the benefits of information age to rural families. The participants were greatly impressed that Swaminathan, always a pioneer in many areas, has yet again become the first scientist to introduce modern

ICT for knowledge empowerment of the rural women and men in general and the resource-poor small and marginal farmers, fishers and landless labour, in particular.

By late 1970 and 1980s, the agricultural extension service in India was becoming progressively weak and even outdated. Consequently, the farmers were not able to get advice on many problems related to crop and animal husbandry. A study by the scientific and technological staff of MSSRF revealed that farmers in general depended on fellow farmers to learn from each other's experiences.

Figure 9.1 explains the weak and strong communication linkages between the farmers and various other agencies. However, this was not always helpful since many of the problems require data and knowledge from the experts in different facets of agriculture. Particularly, advice on management of new diseases and pests, alternate crops necessitated by

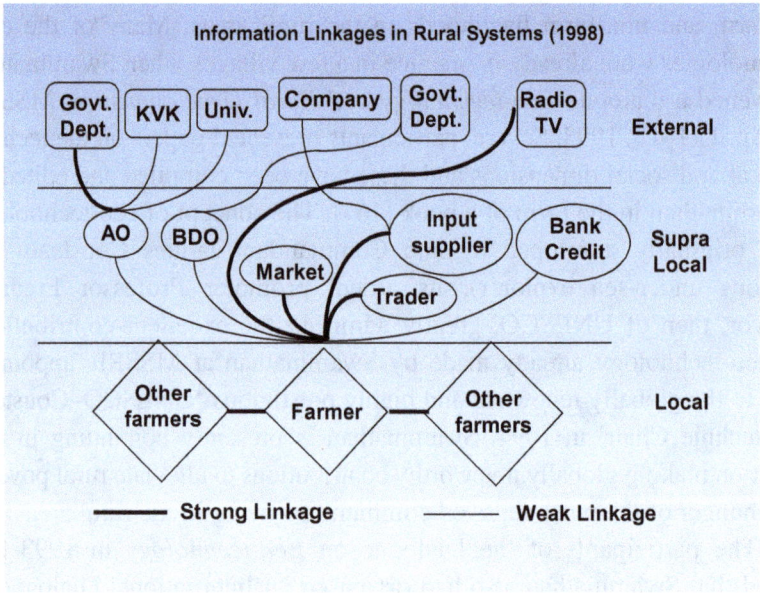

Figure 9.1. Information linkages in rural system, 1998 [2]. Farmer to Farmer linkages — strong; Government officials including Agricultural Officers (AO), Block Development officers (BDO), Bank Credit, etc. — weak; MSSRF's VKCs with Lab to Lab, Lab to Land, Land to Lab and Land to Land Linkages — very strong.

failure of monsoon, as well as climate change was not available to them. Late 1980s was also the period when modern Information and Communication Technology (ICT) was beginning to make impact on the society. Observing the challenges and opportunities, Swaminathan enlisted ICT to link the farmers (data seekers) with experts in various agricultural fields (data holders). So, he went ahead and set up ICT-based information centres in Puducherry. He also chose the term 'knowledge' instead of 'information' since the former is a product of interaction between the rural communities and the scientists/experts, whereas information is passive. He also stipulated that the Village Knowledge Centres (VKCs) in the villages, which wanted them, would be located in common places, so no member of any community would be excluded. In other words, VKCs promoted inclusiveness. He also made it clear that the information content would be developed keeping the local needs and therefore, it is more participatory and 'bottom-up' not 'top-down'. The VKCs gave preference to training young rural women to acquire special computer operation skills (i.e. internet, PowerPoint, video-conferencing, etc.) besides typing, preparing documents, etc. In addition to helping others replicate VKCs in various villages, Swaminathan also sought to have the collaboration with Indian Space Research Organisation (ISRO) in 2004 to enhance the services made available to the villagers. The enhancement came in the form of establishment of Lab to Land, Land to Lab, Lab to Lab and Land to Land linkages, as illustrated in Fig. 9.2.

Earlier biographers [4, 5] have written about Swaminathan's initiative in setting up 'Krishi Vigyan Kendras' (KVKs) to provide institutional mechanism for imparting skill and functional literacy through *techniracy* proposed by him in 1972. They are there in almost all the districts of the country, but unfortunately many of them seem to have lost the dynamic interactions and vibrancy over the years. Swaminathan is of the view that in addition to KVKs, there should be Farm Schools in outstanding farmers' fields. This will promote farmer to farmer learning as it will have high credibility because of trust imposed by farmers on the economics of the farm enterprises of fellow farmers. While farm schools would facilitate farmer to farmer interactions within certain reachable distances, the VKCs' Land to Land can enable farmers from across the country to interact with, of course, the help of interpreters since India is rich in languages. The intangible benefit of

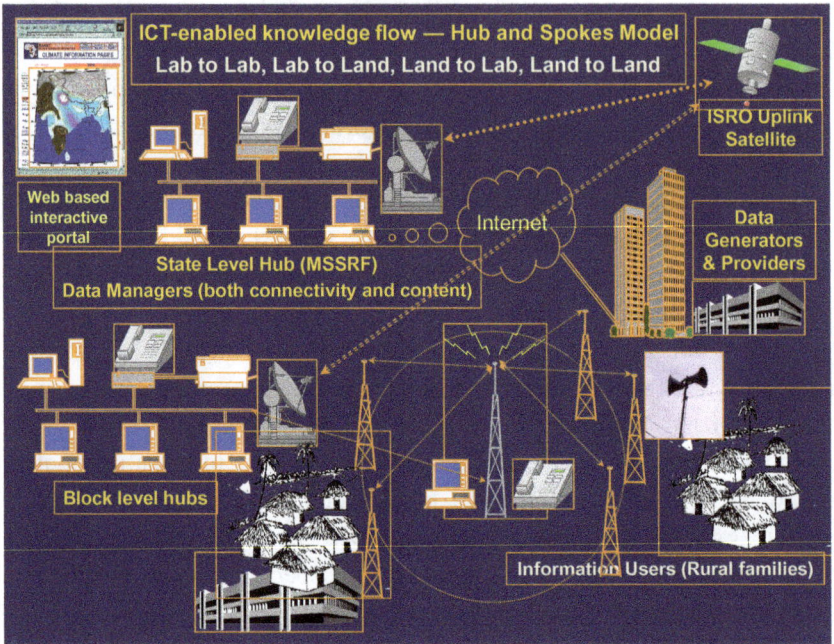

Figure 9.2. ICT-enhanced knowledge flow — Hub and spoke model.

empowering young rural women is that they are no longer subservient to men and that they become reasonably independent to participate in all the decision-making processes at home and the village panchayats. In a nutshell, VKCs bridge both the digital divide and also the gender divide.

The dissemination of locale-specific information/knowledge is almost instantaneous. There are several case studies of farmers saving their crops from pests and diseases with locale and time specific advice from the agricultural research institutions, agricultural universities, KVKs, etc. On a few occasions, pregnant cows caught in serious complication during delivery of calves (i.e. calving) were literally pulled out of the jaws of death only because the distressed women farmers received timely advice through internet from a veterinary doctor located about one hundred or more kilometres away. In 2002, MSSRF won the 'Motorola Gold Award' for its innovative uses of internet in the VKCs of MSSRF with Swaminathan widening the spectrum of the locale-specific information content for the VKCs and VRCs. The VKCs were replicated all over the country both by the Union Government, and the private agencies under the corporate social responsibility (CSR). Earlier in 2001, MSSRF was awarded the 'Stockholm Challenge Award' for effectively bridging the economic and social divide in the rural areas. One of the social divides bridged by MSSRF was to include the excluded low-caste people in the rural areas. For example, a temple that formerly excluded low-caste people now opens its doors to everyone so that they may use the computers placed in the hall of the temple. Thus, MSSRF's VKCs have been the beacon not only in providing the essential knowledge/advisories sought by the resource-poor, small and medium farming, fishing (in coastal villages), and landless rural families, but also in imparting a sense of ownership of the VKCs/VRCs by the communities concerned.

MSSRF's VKCs in several coastal villages of Puducherry provide information on the sea wave heights (which advise the marginal fishers venturing on the sea in their country rafts for fishing far from the shore), the distribution of fish shoals and after fishing in marketing at higher prices by excluding middle men, etc.

On the day of the tsunami (December 26, 2004), the timely warning received from an Indian in Singapore by the VKC in a coastal village (Veerampattinam) in Puducherry was immediately announced on loud-speakers and the villagers were asked to run away from the seashore and rush to the higher elevations. This indeed saved many precious lives especially of children and women.

The use of modern ICT in MSSRF's VKCs/VRCs to obtain the locale-specific, problem-solving knowledge (advice) from various experts working in the various national institutions and universities and disseminating them to the needy women and men has quite substantially and effectively replaced the deteriorating agricultural extension system of the 1960s, 1970s and 1980s. MSSRF's VKCs attracted several top scientists, policy makers, heads of the UN organisations from all over the world. Among them one who took extraordinary interest in the VKCs is Professor Bruce Alberts, an outstanding leader in molecular biology, and also Editor of *Science*, USA. He has visited MSSRF's VKCs about three or four times. During one of such visits, he donated his personal money for MSSRF to set up a VKC. More significantly, he has assiduously explained to the Fellows of the American Academy of Sciences that one of the foreign fellows (i.e. Professor M.S. Swaminathan) of the US Academy of Sciences has made a great innovative application for the modern ICT to establish connectivity among scientists, data holders and data seekers.

During mid-2000, as the Chairman of the National Commission on Farmers (NCFs), Swaminathan made strong recommendation to set up VKCs in each and every one of the 6,38,000 villages of India in order to provide the benefit of 'know-how' and also 'do-how'. In his recent book, *Combating Hunger and Achieving Food Security* [169], Swaminathan writes:

- *Every Panchayat headquarters will have a 'gyan chaupal' or village knowledge centre (VKC). These will have internet connectivity. Alternatively, these can be established in the village school or any public space where there is social inclusion in access to the technology.*
- *The last mile and the last person connectivity will be through FM/ community radio and/or mobile phones. The internet-radio-mobile phone synergy is a powerful tool for social inclusion-in access to all the needed information, including warning of impending natural disasters. Villagers give priority to health and marketing information. In addition, an entitlements database can empower them with information on all the government schemes designed for their well-being. Gender-specific information is equally important. Every farmer in the village should be issued with an 'Entitlements Pass Book'. Artisanal fishermen can now be assisted with a cell phone, which can provide GPS data on wave heights and location of fish shoals.*
- *This programme, VKCs, has evolved into Village Resource Centres (VRCs), and today with satellite communication and mobile telephony it has become a truly transformational technology.*

We recently had a meeting in some of the villages, where women said it has showed them to come out of their homes: *veetlarandhu veliyavanthutom* (in Tamil) — they felt liberated.

Swaminathan's choice of technologies, in fact the eco-technologies are for closing the wide gap in knowledge and skill prevailing between rich and poor nations on the one hand and between rich and poor within all nations. He wrote as early as 1999 [144]: *New information technologies provide a unique opportunity for the knowledge and skill empowerment of the poor. Women and men now living in poverty can be helped to experience a better quality of life only by increasing the economic value of their time and labour.*

One wonders today, in 2016, the great vision Swaminathan had to initiate programmes based on eco-technologies and modern ICT to promote a skill and knowledge empowerment (rather a revolution) in the rural areas. Within the first decade itself, MSSRF's VKCs had brought about a knowledge revolution in the rural areas. Every village in Tamil Nadu, Puducherry and elsewhere used the local bodies (Panchayats) and the government to set up VKCs/VRCs. The women in particular have gained many intangible benefits in addition to enhanced opportunities for rural livelihoods. That knowledge is power and also a powerful tool to free oneself from the shackles of traditional oppression and male domination has now been proven.

Swaminathan who is the UNESCO–Cousteau Ecotechnie Chair was also elected as the 'President of the Pugwash Conferences on Science and World Affairs'. Usually, only the physicists are elected as the Presidents. It is, therefore, a matter of great honour to the biologists all over the world and especially to Indian biologists that Swaminathan was the first ever biologist chosen as the President of the Pugwash. While the founder members Bertrand Russell and Albert Einstein considered human security only from the nuclear weapons point of view, Swaminathan enlarged the concept of 'human security' by adding 'hunger' and pandemics like HIV/AIDS as threats equal to nuclear war. To Swaminathan, the widespread hunger-related deaths as it happened during the Bengal famine of 1943–1944 and the Irish potato famine of 1845 were similar to those caused by 'weapon of mass destruction'. Further, he consistently advocates that 'technology push' should be matched by an 'ethical pull'. He is also of the view 'spiritual globalisation' to accord 'human dignity and gender equity' was needed more than 'economic globalisation'. I believe that as UNESCO–Cousteau Ecotechnie Chair, Swaminathan has enhanced the concept of human security by equating hunger with nuclear weapons and pandemic diseases as weapons of mass destruction. Further, he has also rightly injected the need for ethical and spiritual dimensions in globalisation because without these hunger will continue to remain as weapon of mass destruction.

9.1 Human Resource Development in the Rural Areas

In his book, *Remember Your Humanity — Pathway to Sustainable Food Security* [170], Swaminathan emphasises on Human Resource Development, without which there can be no sustainable development. He points out that MSSRF's institution building philosophy has always been to concentrate on brains and not bricks. The *Gyan Chaupal* movement is stimulating the rural youth to acquire knowledge and skills. *It is equally important that initiatives like VKCs are based on the principle of dynamic and location-specific information delivered in local languages, based on a demand-driven approach*, writes Swaminathan. He stresses that local communities also have a sense of ownership, as otherwise it will not be sustainable. MSSRF's initiative of the Jamsetji Tata National Virtual Academy, which has over 1,500 rural women and men as Fellows as well as 35 foreign Fellows, has become a valuable institutional device to build the self-esteem and capability of rural women and men belonging to socially and economically underprivileged families. External reviewers have recorded that this Academy has helped to convert ordinary people into extraordinary individuals. The National Virtual Academy is another innovation by Swaminathan to recognise and reward the rural women and men with proven traditional knowledge, practical experience in several facets of rural development and strong desire to serve the society. These Fellows constitute a unique entity of human resource with vast practical experience in achieving the goal of sustainable rural development. Often Swaminathan mentions that an ounce of practical experience is worth more than a ton of theoretical knowledge. Rural transformation is unlikely to be brought about by a 'top-down' mode of policy making and implementation by the government. It has got to be 'bottom-up' and participatory with the role of the government mainly being to provide all the necessary support. This is the lesson to be learnt by one and all, based on what has been already demonstrated by Swaminathan.

Chapter 10

Innovator of Strategies to Link Biodiversity Conservation with Livelihood and Food Security

Even the most outstanding cytogeneticists, geneticists and plant breeders have not been visibly active in the conservation of plants, animals and microbes. Hence, it came as a pleasant surprise that Swaminathan, an outstanding cytogeneticist, in his Presidential Address at the XV International Congress of Genetics held at New Delhi in 1983, chose to speak on *Genetic Conservation: Microbes to Man* [123]. This address is one of the papers selected and published in the book *Science and Sustainable Food Security — Selected Papers of M.S. Swaminathan* [171]. Emphasising the need of the biodiversity for the present and for the future well-being of humans and the entire biosphere, Swaminathan in his Presidential Address pleaded for international efforts involving the *ex situ* preservation of germplasm by cryogenic gene banks and *in situ* on-farm conservation.

His plea for a permafrost facility to preserve germplasm of crop plants and their wild relatives was realised in 2008 in the form of *Svalbard Global Seed Vault* set by Norwegian Ministry of Agriculture, but managed jointly by Norwegian Ministry of Agriculture, the Global Crop Diversity Trust and the Nordic Genetic Resource Centre — located in the village of Longyearbyen on Svalbard island situated at 78° north in Arctic circle; an ice mountain was chiselled to create space to store sample seeds and vegetative propagules of about 4.5 million plant species and varieties. The vaults have a natural temperature of –4°C round the year, which is further lowered to –18°C, the optimal temperature for long-term seed viability. The Norwegian Government organised a seminar in February 2009 on *Frozen Seeds in a Frozen Mountain*; and knowing that Swaminathan was the source of inspiration, the Norwegian Government invited him to deliver a lecture on that occasion. The title of his lecture was *Freezing Seeds: A Humanitarian Issue*, which provided a social dimension to the Svalbard Gene Bank. He also likened the Svalbard *ex situ* preservation of millions of cultivars to Noah's ark to serve as safety net for food security in an era of global warming and climate change. He also followed it with a guest editorial [172] in the prestigious journal *Science*. Almost in parallel, the Defence Research and Development Organisation (DRDO) of India has established a similar seed storage facility under permafrost conditions at Chang La in Ladakh and much elated, Swaminathan wrote that such initiatives need to be fostered and given further impetus.

Swaminathan developed highly innovative approaches to revitalise the 'conservation traditions' of the rural and tribal women in India and all the developing countries; back home, he had organised a cryogenic *ex situ* preservation of landraces and indigenous varieties developed and maintained largely by women belonging to a large number of different tribal communities. On a more basic side, Swaminathan elegantly brought out the distinction between the conservation domains receiving public support, and those which are conserved without any public funding for posterity by the tribal women and men living in poverty of varying intensities. Some of these tribal people exist in abject poverty and 'debt trap'. This is what Swaminathan often referred to as *their service for public good at their personal cost*. In this context, Swaminathan elaborates the conservation continuum as having three links as follows:

In situ →	*In situ* conservation →	*Ex situ*
Conservation habitats	On-farm conservation by rural and tribal communities	Conservation through botanical gardens and cryogenic gene banks

Swaminathan was concerned that while the two ends of this conservation chain (namely *in situ* and *ex situ*) receive support from public funds, *in situ* on-farm conservation by rural and tribal men and women largely remains unrecognised and unrewarded. Yet, this link in the chain is primarily responsible for conservation of valuable intra-specific variability.

Biologically important distinction between 'preservation' and 'conservation' arises from evolutionary point of view. The 'preservation' at very low temperatures involves lowering the metabolic processes of the seeds to the barest minimum and this prolongs their viability. One could use these seeds to raise a crop even after a few years to several years depending upon their genetic constitution and cryogenic preservation quality. The disadvantage with the system of preservation is that these seeds do not have opportunity to undergo natural evolutionary processes such as mutations, natural selection, etc. Hence, the advantage of *in situ on-farm* conservation is that further genetic variation goes on. Consequently, the agro-biodiversity is continually enhanced. The challenge was to link the *in situ on-farm* conservation with livelihood (i.e. income generation) to the tribal farmers-cum conservers. In his book *In Search of Biohappiness* [173], Swaminathan in the Introduction writes, *How can we harness biodiversity for poverty alleviation? Obviously, this can be done only if we convert biodiversity into jobs and income on a sustainable basis.* What has been written is not an action envisaged for the future because he has already put the thoughts into practice. An interesting thing about Swaminathan is that when he has extremely novel yet difficult idea, he first tests it in a small scale and then talks about it if it is useful. He has never shot in the dark, and this is the reason why the outstanding and most influential people in the world not only adore him but also keep in touch with him. So, the work he has already done to link conservation with livelihood are briefly as follows: first, he set up field sites of M.S. Swaminathan Research Foundation (MSSRF) at Kolli Hills (Tamil Nadu), Koraput (Odisha), and in Puthoorvayal near Kalpetta in Wayanad (Kerala). Second, he guided MSSRF's partnership with local communities and government agencies in order to strengthen the *in situ* on-farm conservation by rural and tribal communities.

The Koraput region is a primary centre of origin of rice. There are numerous landraces of rice, often genetically resultants of hybridisation and recombination. Hence, they needed to be purified and multiplied and evaluated either for direct use or as donors of genes for rice crop improvement.

With regard to saving seeds of precious landraces and indigenous varieties, a lesson was learnt after the 'super cyclone' that hit coastal Odisha in 1999. Many precious germplasm was lost when the landraces growing in the fields were completely washed out, and in the following days, without grains in their homes to cook, the tribal women cooked and ate the seeds saved for the next sowing season. So, this resulted in the loss/ extinction of several genetically valuable landraces. Under these circumstances, Swaminathan came up with a brilliant idea of developing seed banks. These are the banks with a difference in that the transaction involves seeds and not money. Called the *village gene–seed–grain banks,* these have been playing a very important role in the conservation of landraces and traditional rice varieties and ensuring their continued cultivation. Later, Swaminathan added fodder and water to this banking system which is illustrated below:

Gene → Seed → Grain → Fodder → water
Bank   Bank   Bank   Bank   Bank

The 'Gene Bank' represents the *in situ* on-farm conservation by the rural and tribal families. The 'Seed Bank' involves conservation of representative samples of seeds of landraces and traditional varieties in proper seed stores that will ensure supply of seeds during drought years.

In addition to *in situ* on-farm seed banks, Swaminathan also developed *'Community Gene Bank (CGB) and Herbarium'* that involves cryogenic preservation (*ex situ*), useful as evidence for getting tribal families reward and recognition under Protection of Plant Varieties and Farmers' Right Act (PPVFR).

'Grain Banks' provide grains for food on loan with a stipulation that the borrowed quantity would have to be returned after the harvest with a small additional quantity as interest. 'Fodder banks' also operate as the Grain Banks. In India, the huge cattle population does not have adequate supply of fodder and forage. The Community 'Water Banks' involves community-centric harvesting of rain water and sharing it in times of need equitably.

Swaminathan believes that several farming communities, particularly tribal women farmers, have played a significant role in identifying morphologically (which includes resistance to any biotic/abiotic stress) different plants and multiplying the seeds of these variants. In fact, but for their role as 'breeders' in addition to those as 'conservers' and 'cultivators', the planet would not be so rich with thousands of landraces and indigenous varieties of crops. While the conservation and selection ethos of the tribals had been going on possibly over thousands of years, it is only Swaminathan who, in the later part of the 20th century, evolved strategies, models and draft public policies to accord recognition and reward to their service. One of his strategies was to promote *participatory breeding* programmes. That is to say that tribal women in partnership with scientists (mainly plant breeders) could improve their landraces through hybridisation and selection and pure line selection to purify the mixtures. With the help of MSSRF scientists, the tribal cultivators of landraces of rice have developed much improved strain of *Kalajeera* (a landrace of rice used on festive occasions to prepare sweet dish called *Kalajeera Kheer)*. The improved landrace is called *Kalinga Kalajeera* which fetches premium price in the market.

Swaminathan adopted a different strategy to link the conservation ethos of the tribals in Kolli Hills in Tamil Nadu with their livelihood security. The tribal population in Kolli Hills had been cultivating and conserving a wide range of millets and medicinal plants. Since there was no market demand for millets, the tribals shifted to more remunerative crops such as tapioca and pineapple. Incidentally, pineapples grown in the Kolli Hills by the tribals are organic by default, that is, no chemical fertilisers and pesticides were ever applied, and therefore, MSSRF was able to get 'organic certificate' by a German Company and also to connect Kolli Hills pineapples with an export agency. This again enhanced income to reduce poverty and enhance access to food of the tribals.

Swaminathan also developed a plan for the minor millets cultivated by the tribal people in the Kolli Hills. There was no market for these minor millets nicknamed 'Coarse Millets'. However, chemical analysis of the macro and micronutrients of these revealed that these underutilised millets are highly nutritious and could be ideal dietary items for people suffering from diabetes-2. The nickname 'coarse millets' was changed to *nutritious millets/nutritious cereals* and MSSRF provided market linkages in urban markets for these once neglected millets. Once the health benefits of these minor millets (Little millet (*Panicum sumatrense*), Italian millet

(*Setaria italica*), Finger millet (*Eleusine coracana*), Kodo millet (*Paspalum scrobiculatum*), and Proso millet (*Panicum miliaceum*)) have been brought into public awareness, the demand for these sharply increased. In fact, the demand exceeded the supply and therefore, the cultivation of these had to be substantially increased. It led to the integration of ecology, nutrition and economics in a mutually reinforcing manner. It also forms a continuum, called the C4 continuum, conceived and operationalised by Swaminathan in all the three biodiversity conservation sites (i.e. Kolli Hills, Kalpetta and Koraput). Mention has already been made about the Kolli Hills and Koraput. This continuum involves providing concurrent attention to *cultivation, conservation, consumption* and *commercialisation* (the 4Cs).

Earlier, it was mentioned that Swaminathan, his wife Mina and their three daughters generously donated their plantation land to set up MSSRF's 'Community Agro-biodiversity Centre' (CAbC) at Kalpetta, Wayanad, Kerala. Also referred to as *God's Own Land,* Kerala is blessed with rich biodiversity in crops like rice, banana, jackfruit, tubers, spices, medicinal plants, medicinal rices such as 'Njavara', coconut, areca nut, coastal halophytes and several medicinal plants. In Kalpetta, MSSRF's CAbC is an excellent example of (i) eco-agriculture and (ii) farming with nature or better farming with landscape. The CAbC cultivates several plantation crops (e.g. coffee, vanilla, black pepper, ginger, turmeric, cardamom, etc.), medicinal plants, tuber crops (*Dioscorea* species), jackfruit and several wild but economically useful tree species (*Syzigium travancorium, Cinnamomum malabatrum*) and farm animals. Swaminathan is for conserving every single plant, animal and microbial species. In evidence, MSSRF's Community Agro-biodiversity Centre conserves hundreds of 'high value' wild diversity, particularly rare, endemic and threatened plant species with several of them having medicinal properties. The inputs are largely organic (i.e. farmyard manure, vermiform compost, biofertilisers, biopesticides, etc.) and most of these biological softwares are produced by landless women. This provides additional income to the tribal women. The '4Cs' substantially link ecology with economics and social and gender equities on the one hand, and the continuum of cultivation, conservation, consumption and commercialisation on the other.

Swaminathan has often emphasised that women play a lead role in biodiversity conservation and sustainable use. Notably, he points out that agro-biodiversity is the result of interaction between cultural diversity and biodiversity and that an important aspect of cultural diversity is culinary diversity.

10.1. Plant Variety Protection and Farmers' Right Act 2001

The Trade Related Aspects of Intellectual Property Rights (TRIPS) stipulates that member countries should provide for the protection of plant varieties by an effective *sui generis* systems or any combination thereof. Thus, each country can develop a suitable system to protect its crop varieties. The crux of the problem is that the developed countries do not recognise and reward the contributions of the primary conservers of genetic variability (i.e. rural and tribal people) in crop plants. However, the enormous variability in the form of numerous landraces of rice, millets and other crops is largely because of the tribal women and men who have been saving and selecting seeds for cultivation under their respective agro-ecological conditions and socio-cultural needs. The germplasm of crops selected and multiplied by them have genes for characters like yield, quality and tolerance to a wide range of biotic and abiotic stresses. As discussed earlier, their selfless service for the future food security goes unrecognised and unrewarded, whereas the professional (qualified with post-graduate degrees in genetics and plant breeding) plant breeders who use their material and information to develop new crop varieties are entitled to get varietal protection, rewards and royalties.

Swaminathan decided to end this unethical dichotomy and introduced along with Professor Pat Mooney the concept of *Farmers' Rights* which was endorsed by the UN Food and Agriculture Organization (FAO). Pat Mooney, a Canadian, is known for his work on the conservation of biodiversity. The goal of Farmers' Rights is to accord recognition to the past, present and ongoing contributions of farm families to genetic resources conservation and enhancement. So, for the first time in India, Swaminathan convened two inter-disciplinary dialogues at the MSSRF in 1994 and

1996. Based on the suggestions arising from these dialogues, Swaminathan formulated a draft *Protection of Plant Varieties and Farmers' Rights Act* (PPVFR). Several features of this draft were incorporated by the Union Ministry of Agriculture in its draft legislation. The novel feature of the draft by Swaminathan is that it recognises the farmers as cultivators of crops or conservers of genetic material or both. Further, the MSSRF draft Act provided for creating a 'Community Gene Fund' from which funds can be provided to the farmer-conservers to strengthen their *in situ on-farm* conservation traditions and such other purposes as decided by the farmers. This Act was passed in 2001 by the Parliament of India and India is the only country to recognise Farmers' Rights.

Swaminathan's contributions to conservation of all biodiversity including the agro-biodiversity have been truly phenomenal. He has played a key role in the formulation of the 'Biological Diversity Act' and its passage in the Parliament in 2002. It establishes the sovereign right of the nation over its biological resources and seeks to promote conservation and sustainable use of the biodiversity while also ensuring equitable sharing of the benefits flowing from its commercialisation with communities. The Act provides institutional framework at each 'panchayat' (i.e. grassroot institution at the village level) level for conservation, maintenance of records on the panchayat biodiversity, providing access to its biological diversity with prior informed consent and deciding the share of benefits arising from commercialisation of its biodiversity and traditional knowledge.

Swaminathan's interventions to strengthen the conservation ethos of the tribal women in Koraput (mainly the landraces of rice) led to the tribal women winning the prestigious 'Equator Initiative Award' at the UN Conference on Sustainable Development at Johannesburg, South Africa in 2002. One can imagine how a few women scientists of MSSRF had to train the tribal women in using dining table, spoons and forks as also using the Western toilet system for her to fly to Johannesburg and receive the Award. This conveys the impact that Swaminathan's intervention in the biodiversity-rich tribal communities has made on the various agencies such as United Nations Environmental Programme (UNEP), International Union for Conservation of Nature and Natural Resources (IUCN) which initiated the 'Equator Initiative Award'. Several of the international awards won by Swaminathan are also for his saving nature and saving livelihoods concurrently. In 2007, at MSSRF's Koraput, tribal women farmers won the 'Genome Saviour Award' of the Plant Varieties Protection and Farmers' Rights' Authority of the Government of India.

In a nutshell, Swaminathan has shown the way to the tribal women of Koraput to convert 'biodiversity hotspots' into 'biodiversity happy spots'.

Long before he started innovating strategies to link the conservation of agro-biodiversity in 'hotspots' with livelihood security of the local tribal and rural communities, he started work to restore the degraded mangrove forests along the east coast of India and then link conservation of mangrove ecosystem with gainful livelihood opportunities for the mangrove forest-dependent local communities. During signing of the *Global Biodiversity Protocol* at the UN Conference on Environment and Development in June 1992 at Rio de Janeiro, the *International Society of Mangrove Ecosystem* (ISME) was established in Okinawa (Japan). It was no surprise at all that Swaminathan was elected as its Founder-President. A significant achievement was the preparation of the charter for mangroves conservation and their sustainable use. Under his supervision, MSSRF undertook a massive programme of restoring the degraded mangrove forests along the east coast of India comprising West Bengal, Odisha, Andhra Pradesh and Tamil Nadu. Swaminathan also formed a Committee for *Joint Mangrove Management* (JMM). This comprises all the stakeholders (mangrove forest-dependent local tribal and rural communities, State Forest Department officials, local non-governmental organisations (NGOs) and State Fisheries Department officials). Since the estuaries provide nutrients for the young fish (fingerling), crabs, prawn, etc., inclusion of the Fisheries department in the JMM is quite essential.

The December 26 tsunami caused terrible devastation to the coastal regions of Tamil Nadu. Yet, quite significantly, the magnitude of loss of lives and livelihoods was substantially reduced wherever mangrove forest restoration by MSSRF had taken place. This made the local coastal communities to protect and conserve the mangrove forest and its ecosystem. Mangrove restoration programme by MSSRF under Swaminathan involved nursery experimentation to assess the germination and growth of different mangrove species and setting up pilot plantations before undertaking large-scale restoration. The reasons for the degradation were assessed and then restoration strategies were developed by MSSRF scientists. In Kolli Hills, following successful implementation of the 4Cs of Swaminathan, the income level of the tribal farmers has substantially improved and they proudly sing, 'Biodiversity is our life'. In the coastal areas, mangrove restoration by MSSRF has saved lives and livelihoods. Soon after the PPVFR Act of 2001 and the Biological Diversity Act 2002 came into existence, Swaminathan's concern was to provide awareness about the new Acts. For example, phrases such as 'prior informed consent' need to be precisely understood by the rural and tribal conservers. Towards this need, he has set up *genome literacy* awareness programmes.

To Swaminathan, biodiversity is an ecological asset at the global levels. The UN Convention on Biological Diversity (CBD) recognises geo-political boundaries, but its conservation needs to be at planetary level. With strong commitment to what may be referred to as *our common biodiversity*, Swaminathan has helped several nations to conserve and enhance their biodiversity alongside eco-agriculture. Swaminathan's contributions to sustainable management and conservation of about one million acres of undisturbed Amazonian prime forest of Iwokrama in Central Guyana brings out the fact that his endeavours encompassed the forests, agriculture and in fact the entire biosphere. In February 1990, Shridath Ramphal, Commonwealth Secretary-General invited Swaminathan to help convert one million acres of rainforest into an implementable international project under Commonwealth auspices to demonstrate the concept of sustainable development and conservation of rainforests. Heading a Commonwealth Team to Guyana, Swaminathan prepared a project proposal that resulted in the formation of the *Iwokrama International Centre for Rainforest Conservation and Sustainable Development*. He was also appointed as the Chairman of the Board of Trustees of this Centre from 1992 to 2000 and he guided the activities.

Swaminathan is quite adept in learning while interacting with people. This is exactly what happened when he visited Guyana in May 1990. There he learnt about an old Amerindian legend: *The sky is held up by trees. If forest disappears, the sky — the roof of the world collapses and nature and man perish together.* Moved by this, he wrote in his book, *From Rio de Janeiro to Johannesburg: Action Today and Not Just Promises for Tomorrow* [132], *To our forefathers, forests were not just carbon sinks but were the very foundations of life.*

Back home in India, Swaminathan saved the evergreen forest, called the 'Silent Valley' (so named due to the absence of chirping cicadas) where it was proposed to set up the Silent Valley Hydroelectric Project. The 'Swaminathan Report' recommended the development of Silent Valley into a *National Rainforest Biosphere Reserve* and it also suggested alternate avenues to generate electricity.

Swaminathan was the President of the IUCN during 1984–1990. One of the several things he did was to establish a Global Network of *Mangrove Genetic Resources Conservation Centre* (MGRCC) to arrest the degradation of mangrove forest ecosystem and restore it.

Swaminathan played a significant role in achieving global recognition for the *Gulf of Mannar Marine Biosphere* (GoMMB). Swaminathan's fascination with the 'Kuttanad Below Sea Level Rice Farming' led to his detailed assessment of its potentials for sustainable agriculture in an era of climate change. His fascination was also due to the traditional knowledge and engineering skills of the local communities who since the late 19th century have been raising paddy crop below sea level by reclaiming land from the sea by polders. The polder system developed by the Dutch is globally known, whereas that of Kuttanad polder system has only recently been known largely due to the initiatives and pursuit of Swaminathan.

The Kuttanad Heritage Agriculture — A Delta Region of about 900 sq km has been recently recognised as a *Globally Important Agricultural Heritage System*. Paddy cultivation and fish culturing provide income and food security. The recognition of the globally important agriculture heritage status necessitates revitalisation of some of the vital conservation and enhancement actions of the past that are relevant to address the concerns of sustainable food security, climate change adaptation and ecosystem services management. Several nations with long coast lines, and island states which are currently threatened with sea level rise on account of global warming are deeply interested in the *Kuttanad Below Sea Level Farming*. Apart from delivering lectures in several places, including Japan on *Kuttanad Below Sea Level Agriculture*, he is also suggesting the inclusion of salt-tolerant plants of food and fodder value (i.e. halophytes) in the coastal system of agriculture.

Chapter 11

Developing Strategies to Ensure Food Security in an Era of Climate Change

11.1 Introduction to Climate Change

Climate is the statistical description in terms of 'means' and 'variability' of key weather parameters over a period of time, usually about 30 years. Today, the climate change that is intensely discussed about, especially with regard to its impact on agriculture, including fisheries represents any change in climate over time, whether due to natural causes and/or as a result of human activities.

During the 18th and 19th centuries, there were major inventions and discoveries, each of these in some way or other had a bearing on climate change. The first of these was the invention of steam engine in 1780 by James Watt that ushered in the era of industrial revolution which led to large-scale burning of fossil fuel to extract energy to drive machines, automobiles, and manufacture the chemical fertiliser ammonium nitrate. All these result in the emission of carbon dioxide, nitrous oxide, etc., which are called the *greenhouse gases* which trap heat in the atmosphere, and this in turn leads to rise in the ambient temperature. That carbon dioxide and water vapour in the atmosphere trap heat (i.e. 'greenhouse effect') was first shown in 1859 by John Tyndall. On the other hand, oxygen and nitrogen do not trap heat. Nearly four decades later in 1896, Nobel Laureate Svante Arrhenius predicted that an increase of atmospheric carbon dioxide from burning fossil fuels would lead to global warming. He was a brilliant scientist who also made calculations to show that with the doubling of the then concentration of CO_2 in the atmosphere (about 260 ppm in 1860), the average temperature of the Earth would increase by about 2–6°C. As of May 2016, the CO_2 concentration in the Earth's atmosphere has reached 407 ppm, and the average temperature in comparison with what was about a century ago has increased by 0.90°C. Should the global temperature continue to horse race, the planet will touch an increase of about 1.5–2°C. There is scientific premise to suggest that as the global average temperature increases by about 1.8–2°C, there will occur a *tipping point*. The tipping point would take the planet into a different and altogether unknown stable equilibrium which could be hostile to the welfare and even possibly the very survival of humans and all other species, mainly the mammals. The 'tipping point' would impact the nitrogen, hydrologic and carbon cycles and monsoon patterns in various ways, not precisely predictable yet. That is why Swaminathan since 1970s has been repeatedly cautioning that agriculture would be most highly vulnerable to climate change, than any other sphere of human activity. The influence of climate change on agriculture including fisheries would be both direct and indirect. The direct impact would be on the growth and productivity of the crops, whereas the indirect impact would be routed through the degradation of the ecological foundations of sustainable agriculture, and steep changes in the intensity and pattern of monsoon onset and distribution.

Much of the climate change that is threatening agriculture is largely human-made and hence referred to as *anthropogenic*. Climate Science studies have revealed that Earth's climate assessed in terms of average temperatures has never been static. In fact, there is a discernible periodicity in the cyclic alteration between very long glacial and much shorter interglacial warmer periods. Radioisotope studies conducted in Dome C of Antarctica by several outstanding climate scientists reveal that the paleoclimate of about 6,50,000 years is dominated by long glacial (~1,00,000 years) and much shorter (~10,000 years) inter-glacial (warmer) periods. These studies involved isotopic analyses of the trapped air samples in the glaciers. The isotopes used were carbon-13 for carbon dioxide and methane, and oxygen-18 and nitrogen-15 for nitric oxides. Based on this cyclic changes, the last warming period, the geological epoch 'Holocene', the relatively warmer period of the past 10–12 millennia ought to have ended, and the next successive cooling period (i.e. the 1,00,000 year glaciation) should have commenced. Surely, this has not happened. Hence, the Nobel Laureate Paul Curtzen has coined the term *Anthropocene* to the present, in many ways human-dominated geological epoch, supplementing the 'Holocene'. Top scientists [174] emphasise: *Effective planetary steward-ship must be achieved quickly, as the moments of Anthropocene threaten to tip the complex Earth System out of cyclic glacial-interglacial pattern during which Homo sapiens have evolved and developed. Without such stew-ardship the Anthropocene threatens to become for humanity a one-way trip to an uncertain future in a new, but very different, state of the Earth System.* The planetary stewardship needs to be collective, inclusive of all the national governments and the influential citizens of the world. It should necessarily encompass all the facets of human activities. Of these human activities, agriculture is of major importance, and in this sphere, Swaminathan is the undisputed steward of food security system, especially of the developing countries in the present era of climate change. Way back in 1973, in his Sardar Patel Memorial Lectures *Our Agriculture Future*, Swaminathan [175] proposed the development of *Drought, Flood and Good Weather Codes*. In fact, he stressed that these should be prepared and kept readily available for being put into practice whenever the need arises. In a *larger context, he has stressed that grain reserves are important for food security, and seed reserves are important for crop security.* This takes us back to the 'Gene–Seed–Grain' bank discussed earlier. He wrote in 2002 in his book, *From Rio to Johannesburg* [132], *I proposed the development of 'Drought' and 'Good Weather' codes in order to minimise the adverse impact of monsoon failure and to maximise the benefits of good monsoon. Such 'Action Compacts' are urgently needed. The village level women and men Climate Managers can be trained to implement the codes. If we take the necessary steps, we will be known as a country, which shapes its agri-culture on the basis of an understanding of climate change, rather than as a country whose agriculture is a gamble in the monsoons.*

About 20 years ago, say in 1998, climate change was not an issue even for casual discussion. The question of its possible adverse effects on agriculture does not seem to have crossed the minds of most people, who would matter. Yet, in 1998, Swaminathan [176] wrote on *El Nino and Monsoon Management.* In 1997, Swaminathan had gone to Japan to deliver a lecture at the Climate Conference. In his lecture, he had drawn attention to sea level rise on account of global warming and how the ingression of sea water would lead to salinisation of coastal aquifers and soil. This in turn would result in salinity stress to coastal crops, particularly paddy, the staple crop of a large population. His brilliant idea was to provide 'genetic shield' to paddy by engineering the 'salt tolerance genes' from mangrove species *Avicennia marina* to cultivated rice varieties. His idea received great appreciation and also financial support from Japanese industrialist and the International Tropical Timber Organisation (ITTO). With this support, he initiated the 'anticipatory research', that is, in anticipation of sea level rise and consequent threat to coastal agriculture. The research he initiated and guided at the MSSRF has already resulted in transgenic rice strains capable of withstanding salinity up to 150 mM or about. Since drought is another abiotic stress that would become more frequent, Swaminathan suggested that drought-resistant genes from *Prosopis juliflora*, a common desert tree, to water-thirsty rice. MSSRF scientists are making progress in this project.

In a country where 'business as usual' is the way of life, there are, however, individuals who voluntarily take initiatives, and when it comes to articulating precisely what needs to be done and how to be done, Swaminathan has already emerged as the leader in this field without parallel. This is also evident from his science-based role immediately after the tsunami on December 26, 2004. That tsunami was a terrible calamity resulting in the loss of hundreds of thousands of lives and livelihoods of millions living in the long coastal areas of Tamil Nadu, Andhra Pradesh, Andaman and Nicobar Islands in India, and those of Sri Lanka, Indonesia, Thailand, etc. The entire country was shaken and for the first several hours paralysis of action overwhelmed. Then the usual post-event relief operations started. The usual occurrence after a major extreme natural event is some relief operations which would slowly taper off, and the entire thing is sooner than later forgotten. What Swaminathan [177] did was to write an article titled, *Beyond Tsunami: An Agenda for Action* which formed the basis for the Tsunami rehabilitation integrating elements of sustainable development (i.e. livelihoods, conservation of environment, bioshield, etc.) with village knowledge centres (VKCs) which with their computer and satellite connectivity could render early warning. His agenda for action outlined the immediate as well as the short and long-term measures that should be taken for providing relief to the affected families and for strengthening the coping capacity of the coastal communities in the case of tsunamis in future, and in fact, any major water and weather-related (i.e. hydrometeorological) disasters. The three dimensions in the rehabilitation programme suggested by him are: *life-support and psychological rehabilitation. This is the immediate one. The Medium term rehabilitation covers ecological aspects such as erecting bioshields, restoration of mangrove forests which reduce the velocity of the tsunami waves and hence save significantly the loss of lives and livelihoods. The Long-term (meaning about five to six years) rehabilitation is to strengthen environmental defence systems, enlarging opportunities for sustainable livelihoods based on a pro-nature, pro-poor and pro-women orientation to technology development and dissemination, and improving the productivity, profitability and sustainability of agriculture and fisheries.* The uniqueness of his article in *The Hindu* [177] is not just its incisive approach, but also the sequence of actions proposed. There is no doubt that those survivors, and the destitute women in particular, needed psychological rehabilitation in the foremost, and in the long run, the coping capacity or the resilience. In order to enhance the resilience, the prerequisites are ecological rehabilitation to erect bioshields and livelihood security to access food.

Swaminathan's article in *The Hindu* was not only timely to provide a 'way forward' to plan and implement the rehabilitation schedule, but also quite significant in view of scientific assessment that in an era of climate change, the hydrometeorological disasters would become not only more frequent but also more devastating with increased destructive potential.

Swaminathan has been steadily accelerating his focus on agricultural research and development needed in an era of climate change in order to ensure sustainable agricultural productivity under increased abiotic and biotic stresses. The word 'accelerating' in the previous sentence denotes that he has been deeply involved in agricultural research in an era of climate change since 1970s. His paper on climate change and sustainable nutrition security jointly with late S.K. Sinha was published in 1991 [178]. Based on the results from research endeavours, Swaminathan in his Address at the World Climate Conference in Geneva in 1989 pointed out the serious implications of a rise of 1–2°C in the mean temperature on crop productivity in South-Asia and Sub-Saharan Africa. An expert team constituted by FAO, in its report submitted in September 2009, supported the findings of Swaminathan in 1991 that for each 1°C rise in mean temperature, wheat yield losses in India are likely to be around six million tonnes per year.

So far as the mitigation is concerned, Swaminathan recommends soil carbon sequestration and building soil carbon banks. This could be done only by the national governments.

Several contributions of Swaminathan in developing climate-resilient agriculture, in a sense, form the response to the 'protocols' (i.e. updates) provided from time to time by the Inter-governmental Panel on Climate Change (IPCC) that was set up in 1988. An unpleasant truth regarding the outcome of the Conference of the Parties (CoP) in all the preceding years is that general agreements are only on paper while the climate change is racing towards the tipping point. Some of these agreements are 'politically binding' but not legally. Particularly for agriculture, there has been no decisive strategy. The situation that prevailed just after the climate conference in Durban in December 2011 (CoP-17) is best described by the *Nature* editorial [179]: *It is clear that the science of climate change and the politics of climate change, which claims to represent it, now inhabit parallel worlds.* What *Nature* wrote in 2011 is still largely true in 2016 — several agreements and commitments have piled up, but the field-level action is not yet evident. The economic goals of trade under the globalisation are aggressively pursued, and there appears no space to accommodate the means of survival of the resource-poor smallholder farmers. Both humanity and planet are now hopelessly at the crossroads.

It is under the above-mentioned scenario that Swaminathan's focus to develop field-based, eco-friendly strategies and models to make the resource-poor small and marginal farms climate change risk-resilient and sustainable from ecological, economic and social dimensions makes a huge difference. Briefly these are as follows:

(a) Gene–Seed–Grain–Fodder–Water Banks.
(b) Below sea level farming, the example of Kuttanad: This will compensate to varying degrees for the loss of arable land due to erosion and submergence.

(c) Sea water ingression following sea level increase could lead to progressive salinisation of soil and aquifers. Swaminathan's anticipated research to transfer 'salinity tolerance genes' from the mangrove species (*A. marina*) *via* the recombinant DNA technology has already resulted in remarkable success at MSSRF. Since the genetically engineered crops are at present controversial, MSSRF has also isolated landraces of rice that are naturally resistant to both salinity and submergence.

The findings of such naturally climate change risk-resilient variants in crop plants absolutely support the many great endeavours of Swaminathan to conserve and enhance agro-biodiversity. His guest editorial [172] *Gene Banks for a Warming Planet* elaborates the importance.

Swaminathan is at present guiding research to: (i) Further transfer the 'salinity-tolerant genes' of *Avicennia* already engineered to a few rice strains (i.e. an outcome of strategic and anticipatory research) to the locally adapted landraces/indigenous varieties of rice through 'participatory' breeding involving the rural/tribal people and professional plant breeders. This would also help the resource-poor small and marginal farmers to derive benefit from the PPVFR Act-2001, (ii) Utilise new agronomic practices such as the System of Rice Intensification (SRI) which is a sort of *dry rice cultivation*. In this system, the paddy remains wet, but not saturated as in water-logging conditions. The partial stress induced in the rice plants activates the physiology and brings about a plethora of favourable effects such as deep-penetrating root system, increased efficiency in absorbing nitrogen, hardier plant, and greater resistance to pests and diseases. The yield is also significantly increased because the tillering is profuse, thereby producing more grains per plant. Under water-stress conditions, SRI is the solution to achieve appreciable productivity. Swaminathan observes that if this method is applied to 20 million hectares of land under rice cultivation in India, adequate rice for the country could be produced using 'more crops per drop of water' technique. (iii) Swaminathan [9] has made clear-cut prescriptions for climate-resilient management of agriculture. The following are the components:

(a) conservation farming and climate-resilient agriculture in the heartland of Green Revolution, viz., Punjab, Haryana and Western UP in order to defend the gains already made;

(b) bridging the gap between potential and actual yields in Eastern India through integrated packages of technology, services and public policies;

(c) launching a pulses and oilseeds revolution in rain fed areas through organisation of pulses and oilseeds villages;

(d) launching a post-harvest technology and value-addition revolution by ending the prevailing mismatch between production and post-harvest technologies and by promoting agro-processing and agribusiness;

(e) bringing about a management revolution leading to equipping small and marginal farmers with the power and economy of scale.

Swaminathan has greatly emphasised the absolute need for conservation of biodiversity in a 'warming' planet. This aspect ranging from preservation of seeds under permafrost conditions to conservation by *in-situ* on-farm system has already been discussed. Swaminathan [172] points out, *biodiversity enables the development of plant varieties with novel genetic combinations, which will be required to meet the challenges arising from adverse alterations in temperature, precipitation and sea level as well as more frequent droughts and floods all of which are anticipated from human-induced climate change.* In this regard, reference has been made earlier to Svalbard Gene Valut in Norway and Chang La in Ladakh, India.

Community Management of Climate Change especially in the rural areas is strongly recommended by Swaminathan. The MSSRF has initiated a programme for training Community Climate Risk Managers. MSSRF's VKCs and Village Resource Centres (VRCs) which provide locale-specific, demand-driven information are rapidly generating and/or obtaining weather and crop-related data; in addition, these are providing necessary training and capacity building for the 'Community Climate Risk Managers'. Such managers will be familiar with the art and science of managing climate risks like drought, flood, higher temperature and sea level rise. The climate risk managers are trained to manage the 'Good Weather Code, Drought Code, Flood Code', etc.

Swaminathan has also considered the need of energy component in the sustainability of agriculture, especially of resource-poor small and marginal farms. The off-grid villages do not contribute to the emission of greenhouse gases to any appreciable levels. Yet, there is the need of energy in one form or another to run machinery, irrigation, and small-scale village industries. Without energy, the agricultural activities frequently remain at a subsistence level. In the off-grid villages, the small holder farmers have to mostly rely on farm animals for draught purposes, supplemented with the labour of rural women and men belonging to family farms as well as landless families. An off-grid smart village is one that achieves food security by using pathways of production that depend increasingly on biological rather than chemical inputs. These reduce the need for energy-dependent chemical fertilisers; in turn, this results in reducing emission of greenhouse gases. In an earlier chapter, dealing with Evergreen Revolution and eco-technologies driven 'on-farm and non-farm eco-enterprises', reference was made to provision of livelihoods to the landless rural women who use their manual energy largely to produce vermicompost, green manure, biofertilisers and biopesticides. The pro-nature, pro-poor, pro-women and pro-livelihoods dimensions of these also represent 'pro-mitigation of climate change'.

11.2 Using Nature's Gifts

In addition to the technology-based human efforts to genetically shield the coastal agricultural crops against salinity stress, there are also spontaneously occurring crop varieties with salinity tolerance, drought tolerance, etc. Halophytes (i.e. plants which grow well in sea water) are the gifts of nature — a consequence of spontaneous mutations and natural selection favouring survival in sea water. These variants are either directly used in cultivation or else chosen as 'donors' of salinity-tolerant genes for genetic shielding of the locally adapted high-yielding crop varieties. These used directly or in crosses with high-yielding varieties to produce desirable segregants would be free from the public concerns related to transgenic crops and so, these would be readily acceptable for human consumption at home and abroad. The food, fodder and fibre crops which are capable of growing and producing under high salinity conditions are the 'halophytes'. Swaminathan, through MSSRF, has set up a 'Genetic Garden of Halophytes' in Vedaranyam in Tamil Nadu, India.

From the foregoing, it is obvious that there is no single approach/ pathway to sustain agriculture both to produce food and rural livelihoods under the threat of climate change risks manifested as higher temperature, floods, submergence, prolonged droughts and extreme hydrometeorological events. Therefore, Swaminathan has developed a variety of strategies and models to safeguard the agriculture especially of the resource-poor smallholder farms which constitute hundreds of millions in India.

I have not known any agricultural scientist except Swaminathan who dealt with climate change risks and their management as early as 1973. In his Sardar Patel Memorial Lectures on *Our Agricultural Future* [175], Swaminathan dealt at length with the adverse impact of emerging climate change on agriculture and what could be done to mitigate or even overcome some of them.

Under Swaminathan's direction, MSSRF is the only institution that has transgenic crop (rice) to tolerate a climate change-induced rigorous stress, namely salinity and drought. The genetically modified crops produced by multinational companies are pesticide-producing either Bt — against the borer pests (e.g. *Helicoverpa armigera*) or the herbicide-tolerant (Ht) crops which can survive application of high doses of herbicide, 'Roundup' with active principle being glyphosate. Glyphosate has been recently declared as '2A' carcinogen by the International Agency for Research on Cancer (IARC). The fundamental evolutionary principle, namely, development of genetic resistance by the pests under 'selection pressure' is ignored. As of now, the pests have already acquired resistance necessitating application of larger and larger quantities of pesticides, the detrimental health and environmental consequences of which are absolutely indisputable. What drives this situation is the motive to make money at any cost in a globalised world. In his Report of 2004, as the Chairman of the Task Force on Agricultural Biotechnology, Swaminathan had recommended that transgenic crops producing pesticides biosynthetically in their system should, if at all, be accorded the lowest priority. However, today, these are the only ones either already commercialised or under commercialisation, since these would ensure continued production of pesticides alongside the transgenic crops. Needless to overemphasise, the target is not to make crops adapt to climate change risks, or to develop strategies to minimise its detrimental effects, but to develop genetically engineered crops that are relatively easy to develop and make huge profits.

The future generations will judge the contributions of Swaminathan which are basically pro-ecology and pro-poor *vis-a-vis* those which are injurious to the ecological foundations of sustainable agriculture, rich–poor divide, and gender inequities in developing strategies and models to mitigate/overcome the climate challenge risks. Swaminathan and Kesavan [180] in their paper, *Agricultural Research in an Era of Climate Change* have discussed the nature of basic research of interdisciplinary approach required to achieve better control over the adverse impact of higher temperature and higher concentration of atmospheric CO_2 on the growth and productivity of major crop plants. There are already several reports that increased CO_2 concentration in the atmosphere increases photosynthetic efficiency and biomass of crop plants, forest trees and several other plant species. There is a general agreement that the C3 plants such as wheat, barley, rice and potatoes respond positively to CO_2 enrichment. However, as has been shown by Sinha and Swaminathan [178], the rise in temperature reduces the yield by reducing crop duration. Therefore, critical experimental studies are needed to understand the effects of higher temperature and higher CO_2 concentrations on the growth and productivity of the C3 crop plants.

In conclusion, climate change risks are clearly anthropogenic; it is the price that humankind has to pay for its greed, social and gender inequities and neglect of ecology and degradation of environmental resources. It is sad but true that as of now, the politics of climate change has no solution to retard its 'Horse Race' towards 'tipping point'.

Chapter 12

In Quest of a World Without Hunger

Several biographies of Swaminathan especially those written by Iyer [4] and Dil [1] indicate that Swaminathan's resolve to eliminate hunger from the face of Earth came to be imprinted on his mind when he was in his late teens coinciding with the Bengal famine of 1943–1944. Basically a Gandhian himself, hunger-related deaths were unbearable to him. Further, that famine-related hunger could strike anywhere in the world was evident from the Irish potato famine of 1845. Food is the most basic among a hierarchy of human needs. His firm resolve to study agriculture and solve the world's hunger problem even made him overcome his sentiment to study medicine and take over his late father's hospital in Kumbakonam. Hence, his quest for a hunger-free world started about seven decades ago.

As described in Chapter 2, it was an unlikely coincidence that the first crop for his research in Wageningen happened to be potato — the crop which was devastated by a fungal disease, 'Late blight' to result in famine; to more spiritually oriented minds, it would seem that cytogenetic research in potato was a divine ordain to Swaminathan.

It is now amply clear that solving the hunger problem in the world cannot be achieved by just increasing food production. When one refers to ensuring food security to eliminate hunger at the individual level, the following three major requirements/components must be addressed:

- **Availability** of food in the market, which is a function of production in the country and/or importation. The Irish potato famine was the result of failure of potato crop due to a fungal disease (*Phytophthora infestans* — Late blight).

- **Access** to food, which is a function of adequate purchasing power (i.e. a function of employment/livelihoods). This is a significant factor for the large number of hungry people in India.
- **Absorption** of food in the human system, which is a function of clean drinking water, sanitation, primary health care, etc.

These three components of food security are important for India for the following reasons:

(i) In the USA, the average farm size is about 200 ha, whereas in India, about 80% of over 115 million, farming families are small (2 ha and less) and marginal (1 ha and less).

(ii) In the USA, less than 2% of the population is engaged in farming and they produce more than enough food for the entire population of 314 million people; so, only about 2% of the US population are the 'farmer-consumers' — a term used by Swaminathan in his book [170] *Remember Your Humanity — Pathway to Sustainable Food Security*. On the other hand, in India, over 60% of the population of about 1,300 million people are 'farmer-consumers'. Most of them practising the subsistence farming do not have adequate marketable surplus, and without money, they remain in abject poverty and often fall into 'debt trap'.

Today, the world has enough food to feed the entire global population of about 7.5 billion. But about a billion people go to bed hungry. India alone has more than 25% of the world's hungry people. The reason is lack of employment/livelihoods leaves them poor without money to access food. Globalisation has not been kind towards the jobless; corporate sector, both industrial and agricultural (i.e. factory farming), has been eliminating nearly two-third of jobs in order to achieve greater profits through 'jobless economic growth'. So, increase in the number of hungry people is due to lack of money and not food. India, therefore, should provide all the necessary support to promote smallholder family farms and transform subsistence farming into dynamic agri-business units. In his book, *In search of Biohappiness,* Swaminathan writes [173]: *Although the Indian Government has more than 50 million tons of wheat and rice in its warehouses, over 250 million women, children and men still go to bed hungry. Therefore, jobs should be the bottom line of our agricultural policy. Where there is work, there is money, and where there is money there is food.* This is so far as 'access' to food is concerned.

Swaminathan's concept and resolve to achieve a 'Zero Hunger' state goes beyond providing cereal-based food to quench the hunger fully or partially. To him, it has to be a balanced diet. Writing on *Achieving the Goal of Food Security and Nutrition for All* [181], Swaminathan first provides data to describe India's appalling position in regard to its performance in the Millennium Development Goals (MDGs) that ended in December 2015 and Global Hunger Index (GHI). He writes, *The National Family Health Survey 3 showed that as of 2002–2006, 45% of children in India under 3 years of age were stunted and undernourished. One in 3 undernourished children lives in India and undernutrition is associated with higher under-5 deaths. Long term effects of early undernutrition include cognitive and physical growth deficits across multiple generations and reduction in immunity to infections.*

India's position in the GHI is 55th among 76 countries. Sri Lanka, Pakistan, Bangladesh have slightly better ranking than India. The GHI has been developed by the International Food Policy Research Institute (IFPRI), and it is calculated on the basis of the following criteria:

- **Undernourishment:** the proportion of undernourished people is a percentage of the population (reflecting the proportion of the population without adequate food, i.e. insufficient calorie intake);
- **Underweight children:** the proportion of children under the age of five who are underweight (i.e. they have low weight for their age, reflecting wasting, stunted growth, or both) which is a widely used indicator of child undernutrition;
- **Child mortality:** the mortality rate of children under the age of five (partially reflecting the often fatal synergy of inadequate food intake and unhealthy environments).

Levels of children being underweight in India is at 43% (twice the average level of 21% reported in sub-Saharan Africa), and stunting is at 48% (8% higher than that reported in sub-Saharan Africa).

It is not that children alone suffer from inadequate food intake. Today, about 1 billion out of about 7.5 billion people in the world go to bed unfed or underfed. This is only with reference to under or malnutrition. In his recent book, *Combating Hunger and Achieving Food Security* [169], Swaminathan writes that nutritional maladies may take the following forms:

- **Protein energy under or malnutrition**, primarily caused by poverty-induced lack of purchasing power — Swaminathan's prescriptions include ethics in the Public Distribution System, eco-technologies-driven rural eco-enterprises, and legal Entitlement to food, in addition to public welfare schemes of the Government such as 'Mahatma Gandhi National Rural Employment Guarantee Act' (MGNREGA), etc.
- **Hidden hunger** caused by the deficiency of micronutrients in the diet, such as deficiency of iron, iodine, zinc, vitamin A, vitamin B12, etc.
- **Transient hunger** arising from either natural calamities or civil disturbances, including ethnic conflicts.

As early as 1985, Swaminathan [182] clearly articulated the three evolutionary steps to be adhered to in dealing with food systems. These are as follows:

(i) **Food Self-sufficiency** which implies adequate supplies in the market.

(ii) **Food Security** which involves both physical and economic access to food.

(iii) **Nutrition Security** which implies physical and economic access to balanced diets and safe drinking water for all people at all times. In the context of climate change, Swaminathan observes [183] that the progress of food systems is rather slow and inadequate. He states, *Many developing countries including India which have experienced a Green Revolution are still in the first stage of this evolutionary process.*

Most of these countries, especially, India with over 600 million 'farmer-consumers' are grappling with the famine of rural livelihoods.

With inflation and escalation in the prices of fruits and vegetables, coupled with limited or no access to these, the 'hidden hunger' is engulfing increasingly large proportion of the population, especially in India and a few other developing countries. Even if hidden hunger (caused by the deficiency of micronutrients such as iron, iodine, zinc, vitamin A, vitamin B12, vitamin C, vitamin D, etc.) is the main cause of debilitation morbidity and death, particularly among the economically weaker sections of the society, it could be mistaken as due to undernutrition. This is so because, as the term implies, the hidden hunger caused by micronutrient deficiency is not readily discernible. In this regard, Swaminathan's focus on the hidden hunger and the development, in the recent years, of strategies and models to combat it became socially relevant and globally distinct. The distinctness of Swaminathan's approach is assessed from the novel and readily 'doable' proposals and action plans developed by him. These are as follows:

The UN Secretary-General Ban Ki-Moon launched at the UN Conference on Environment and Development in Rio de Janeiro in June 2012, a *Zero Hunger Challenge* with a target date of 2025 for the elimination of hunger, malnutrition and food insecurity at the individual household levels of hundreds of millions of people around the world. The five pillars of the *Zero Hunger Challenge* are:

(i) 100% access to adequate food all year round;
(ii) zero stunted children less than two years of age;
(iii) 100% increase in smallholder productivity and income;
(iv) all food systems to be sustainable;
(v) zero loss or waste of food.

The mechanisms of realising or erecting these five pillars need to be outlined. It is in this regard that Swaminathan wasted no time at all in moving ahead with speed and effectiveness to come up with ideas and action plans to make use of the millions of resource-poor small and marginal farms to tackle the *hidden* hunger along with *protein* and *caloric* hunger.

An immediate step that Swaminathan took soon after the declaration by Ban Ki-Moon's *Zero Hunger Challenge* was to write an editorial in *Science* [184] under the title *Combating Hunger*. In this editorial, he refers to the High Level Panel of Experts (HLPE) to the United Nations on World Food Security that was chaired by him. The Panel released a comprehensive report on *Social Protection for Food Security* to combat chronic childhood hunger. The HLPE chaired by Swaminathan proposed the concept of *Food Security Floor* and is the adoption of a bottom-line approach in public policies related to achieving the goal of sustainable food security for all in every country. The *food security floor* will indicate the minimum steps needed for ensuring that every child, woman and man will have an opportunity for a healthy and productive life through access to balanced diet, clean drinking water, sanitation and primary health care. The basic philosophy of the *food security floor* is that freedom from hunger is a fundamental human right defining the minimal steps such as those mentioned above. The UN designated 2014 as the 'International Year of Family Farming' recognising that about 500 million family farms of varying sizes, involving over two billion people, play a key role in food production and food security worldwide. Writing yet another editorial in *Science* under the title *Zero Hunger*, Swaminathan [185] has suggested that family farming, characterised by diverse crops, can be harnessed to support nutrition-sensitive agriculture. Noting that commercial farming tends to promote market-driven monoculture of food crops, which ignore nutrient needs, the family farms, both small holder as well as bigger ones can cultivate diverse crops with nutrient content specific to eliminate a deficiency.

Devoting much of his time in the recent past, Swaminathan has been constantly refining his approach to providing agro-horticultural remedies to nutritional maladies. These ought to be 'doable' in the resource-poor small and marginal farms in pro-nature, pro-poor, pro-women and pro-livelihood generation manner. His recent proposal of a farming system model to leverage agriculture for nutritional outcomes, called 'Farming System for Nutrition' (FSN) comprehensively fulfils not only the pro-nature, pro-poor, pro-women and pro-livelihood orientations to farming, but also the three dimensions, viz., ecology, economics and equity of sustainable agriculture and rural development. It is perhaps the most simple in design, yet highly effective to provide agri-horticultural remedies to nutritional maladies in a given agro-ecological zone. He calls it the nutrition-sensitive agriculture [186].

In a nutshell, Swaminathan's strategies to overcome the three major forms of hunger, viz., 'undernutrition', 'protein hunger' and 'hidden hunger' are nationally and internationally acclaimed. Of course, livelihoods to access food of specific kinds to eliminate all the three kinds of hunger is the major prerequisite. In this regard, Swaminathan's strategy of knowledge and skill empowerment of the rural and tribal people through *techniracy* in the biovillages and VKCs has already been elaborated.

With regard to availability, India now has adequate quantities of cereal grains. There is, however, shortage of production of pulses especially for the predominantly vegetarian section of the people. In order to overcome protein hunger, Swaminathan has shown the effectiveness of promoting the concept of 'Pulse Panchayats' to encourage cultivation and consumption of pulses in villages. In his paper *Achieving the Goal of Food Security and Nutrition for All*, Swaminathan [181] provides a list of horticultural plants such as spinach, watermelon, mushrooms, guava, lemon, sweet potato, lettuce, dark leafy vegetables to remedy nutritional maladies caused by deficiencies of iron, calcium, folic acid, vitamin C, vitamin A, etc.

He calls for a food-based approach to solve the 'hidden hunger' than a drug or tablet-based treatment. He recommends that diets should include biofortified plants like moringa, sweet potato, pomegranate, nutri-millets and fruits and vegetables besides milk, eggs, etc.

To the plant breeders, Swaminathan suggests developing biofortified varieties of crop plants by breeding plants such as iron-rich pearl millet, zinc-rich rice, etc.

Further, he recommends development of a cadre of 'Community Hunger Fighters' to master the science and art of leveraging agriculture for nutrition. Swaminathan also recommends that the existing kitchen gardens should be redesignated as 'Nutrition Garden' growing spinach, banana, cabbage, tomato, sweet potato, beetroot, carrot, lettuce, etc.

Swaminathan's innovative approach for achieving sustainable food security to eliminate caloric, protein and hidden hunger involves the following:

(i) Evergreen Revolution to comprehensively include eco-agriculture and eco-enterprises to fight the famines of both 'food availability' and 'rural livelihoods' to access food.

(ii) The FSN integrated within the domain of eco-agriculture involves location-specific inclusive models to address nutritional needs of farm and non-farm families in a given agro-climatic zone. The main components of the FSN are as follows:

    (a) surveying and identifying the major nutritional maladies;

    (b) designing suitable agricultural interventions to address the problems;

    (c) including specific nutritional criteria in the design;

    (d) improving small farm productivity and profitability;

    (e) undertaking nutrition awareness programmes;

    (f) introducing monitoring systems for assessing impact on nutrition outcomes.

All these emphasise that nutrition-sensitive agriculture can be established in the resource-poor small and medium farms. Many of them are smallholder family farms, and operationally the FSN is pro-nature, pro-poor, pro-women and pro-livelihoods.

Finally, the FSN is absolutely an innovation in India by an Indian, Swaminathan. Its roots are traditional. Every aspect of the FSN is in harmony with nature. What is, however, most important about the FSN is its favourable economic, social and ecological attributes in comparison with products of modern biotechnology. Swaminathan is certainly not divorced from any technology including the modern agricultural biotechnology, but he favours those which are eco-friendly without any doubt and which in the long run would not become the weapons of destruction (i.e. hatchet).

Finally, it must be acknowledged by one and all in India and the entire humanity that Swaminathan is truly a legend in science simply for the reason that he has piloted the journey from the days of acute food shortage until the mid-1960s to the current destination of *food for all the Indians.*

The transition from Bengal famine to Right to Food with home-grown food from the days of ship-to-mouth to shipping food is incredible but true. Gandhiji, greatly dismayed by the Bengal famine of 1943–1944, expressed his anguish by saying that 'God could appear before the hungry only in the form of bread'. He knew hunger is the most cruel form of physical suffering. Swaminathan was a student then. Today, Swaminathan has made God appear before the hungry and destitute in the form of bread. Further, his current endeavours are to add proteins and micronutrients to the bread. Gandhiji fell a martyr to an assassin's bullet in January 1948, but it is only now that his soul would rest in peace after the greatest son of India, Swaminathan, has fulfilled Mahatma's prayers that no one in independent India should feel the pangs of hunger — the most cruel form of physical suffering of all living beings including the humans.

To Mahatma Gandhi, freedom from hunger meant lot more than freedom from the colonial rule; so, if he were to be alive today, his happiness that Swaminathan, son of his close friend, late Sambasivan of Kumbakonam, has eliminated the pangs of hunger, especially of the rural poor, from independent India, would have risen to boundless heights. He would approve that Swaminathan is truly a 'Legend in Science and Beyond'.

Epilogue: The author (PCK) in conversation with the 'Legend and beyond in Science' (MSS)

PCK: This conversation with you is for the epilogue of the biography entitled 'M.S. Swaminathan: Legend and beyond in Science'. In this biography, I have made analyses of your journey in science and its application for food security over the past 65 years. Your scientific contributions bring out excellence in areas of contemporary relevance. The question is how you could have so easily made the transition from highly basic research far ahead of the times, then to applied research for overcoming various hurdles and erasing the image of India as 'begging bowl' to 'bread basket'. Your views and experience in this regard would be very useful for our youth.

MSS: I have always considered science as a continuation with basic, applied and translational research having a feedback relationship. As I started my research career the agenda became rather wide but I had always kept some applied objective in mind. For example, my interest in induced mutagenesis started when I needed dwarf plant architecture in wheat and rice. I had then read Dr A Gustafsson's paper on the erectoides mutants in barley and wheat. I invited Professor Gustafsson to IARI and his lectures were fascinating with reference to the application of induced mutagenesis in crop and animal improvement. I also spent a week in his laboratory in Stockholm and learnt the application of X-rays in studying seed fertility. Day to day work resulted in many questions and also helped to identify areas which need in-depth examination. This is why I got involved in a variety of issues including crop yields, polyploidy and radiation sensitiv-ity. You have cited other examples of this curiosity-led research. At the same time I used to travel to villages and saw the poor economic condi-tions of our farmers as a result of poor agricultural productivity. As a scientist I have an obligation to help farmers come out from the low yield and low income traps.

PCK: Globalisation has not benefited India; further, the economic thrust in the trade under globalisation is forcing nations to ignore the environ-mental and social (and gender) dimensions of sustainable development. Intra- and inter-generational inequalities within and among the nations are widening rapidly. The present era is characterised by undesirable global change in economic, environmental and social spheres of sustainable development. What are your views on this complex situation?

MSS: Globalisation will confer benefits if it leads to common goals such as eradicating poverty, malnutrition and environmental degradation. It should not be based only on trade competition and making money. Environmental protection, of which reduction in emission of greenhouse gases is an important component, would lead to benefits for all. Globalisation will be successful only when there is a win-win situation for all stakeholders. Conservation of biodiversity and the effective implemen-tation of the Global Biodiversity Convention will help to promote an era of biohappiness. Biohappiness involves the sustainable conversion of biodiversity into jobs and income. The present era of global change has both positive and negative consequences. It is said that we are entering the age of Anthropocene where human action will determine whether we are going to lead the human family to a happy future or to environmental disasters. We should try to make Anthropocene the beginning of a global partnership for Gross National Happiness (GNH) rather than GDP as a better measure of the impact of human activities on the environment.

Today, our scientific and academic infrastructure is really good and there is also more government support for science and technology. Therefore, one could take up problems now which are very different from the one I took up in the 50s and 60s. My main consideration at the time was self-sufficiency in food and eradication of hunger. Partly this goal has been achieved but we are still confronted with the challenge of hunger and malnutrition. This is why I have continued to accord priority in my own work to overcome the wide spread malnutrition prevailing in the country. My latest contributions in this area are organization of Farming System for Nutrition (FSN) and the establishment of 'Genetic Gardens of biofortified crops'. Biofortification which I have been promoting helps to marry nutrition and agriculture. This will also provide livelihood opportunities to millions of women and men who still go to bed hungry. We cannot be silent to a situation in which hundreds of millions of people have no food security for want of access to food. This explains my current priority in the allocation of my time.

PCK: Way back in 1974, well over four decades ago, you wrote about the possible adverse impact of climate change risks on agriculture including fisheries. No one had ever dreamt then of climate change. While most of us cannot see such major planetary level changes about to occur even within a few years from now, how were you able to foresee such things several decades earlier?

MSS: Having seen that the disastrous effects of drought and flood in the 1950s and 60s, I pointed out in my Sardar Patel Lectures of the All India Radio in 1974 that we must take anticipatory action to mitigate the adverse impact of aberrant weather. For this purpose, I proposed a set of action points grouped under generic terms such as drought code, flood code and good weather code. In 1978, I was invited to deliver a plenary lecture at the World Meteorological Conference on the topic 'Climate and Agriculture'. In 1988, I was invited by the World Meteorological Organization to deliver a public lecture on 'Climate Change and Agriculture'. Within a span of ten years climate change became important. Unfortunately the various preparatory action I recommended were not acted upon with the result that we are facing extremely difficult circumstances in agriculture including drinking water supplies. At least now there is more awareness of the consequences of climate change. At MSSRF we started working in 1990 on the mitigation of sea level rise. Conservation of mangrove helped to reduce the fury of flood during the titanic tsunami of 2004. We should also be prepared for an increase in average temperature by about 2° C. Such an increase in mean temperature will result in reducing the duration of the crop. About 6 million tonnes will be lost in wheat.

PCK: Professor Jeffrey Sachs of the Earth Institute, Colombia University has observed that the Green Revolution has come at huge and irreversible environmental costs. He does not refer to your Evergreen Revolution which is largely sustainable and Professor E.O. Wilson of Harvard University has applauded it in his epoch-making book *The future of life* (2002, The Vintage Books, London). In fact, you have referred to the new dwarf and semi-dwarf wheat–based agriculture (i.e. chemically intensive) as 'exploitative' and that it should be used only as for 'breathing space'. Your statement might have been largely unwelcome and even annoying to several people. Yet, you went ahead. Your courage and conviction, in my view, add substantially to the three pillars viz. ecological, social and economic pillars of sustainable development. I would say 'courage and conviction' could well be the fourth dimension of sustainable development. Do you agree with this view? If so, how would you like to put it across to the well-meaning youth who take up job with a 'mission', and not just for money?

MSS: When the term Green Revolution was coined in 1968 by Dr William Gaud of the United States, I mentioned that this should not become exploitative agriculture. On the other hand Green Revolution means more crop per unit of land and water. We are observing shrinkage in per capita availability of water and land. Therefore we have no option except to produce more food and other agricultural commodities from less land and water, but this needs to be achieved without harm to the environment. For this purpose we need to develop technologies, which can help to increase productivity in perpetuity. This is what I termed as 'Evergreen Revolution'. Evergreen Revolution implies the mainstreaming of the principles of ecology in technology development and dissemination. We shall develop biological software for ecological agriculture. I am glad these ideas are now getting widely accepted. I agree that courage and conviction are important requirements for sustainable development. We need to promote green agriculture which is environment friendly through the adoption of integrated pest management, integrated nutrient supply, etc. Evergreen Revolution has to be a mission and should become an important part of Goal 2 of the UN Sustainable Development Goals. This will be the right beginning for anthropocene.

PCK: Today, the term 'anthropocene' encompasses the functioning of the entire earth systems. The Haber-Weiss chemical process effectively converts atmospheric nitrogen into ammonia, but there is as yet no chemical process for converting the accumulating nitrates in the soils and aquifers back into nitrogen. So, we have essentially developed a one-way processing system with respect to thousands of human-made chemical substances that nature is unable to readily digest and recycle. So, the point is that environmental stresses to crop plants, farm animals and edible marine resources are arising from degradation of what you refer to as the 'ecological foundations', besides the climate change risks. A plausible argument could be that looking at only climate change risks without addressing the other changes in the functioning of the Earth systems will not do. If you agree with these, what is your advice to save agriculture and produce enough to feed the burgeoning human population? Is 'Evergreen Revolution' adequately capable or something else needed as supplement to Evergreen Revolution?

MSS: Anthropocene has to be viewed in a holistic way. It requires a systems approach to the management of natural resources. It should be our endeavour to convert Anthropocene into a developmental pathway which will enhance the quality of life of current and future generations. As I mentioned earlier we will have less per capita availability of land and water in the future. Green agriculture, based on eco-development will be essential to enter into an era of Evergreen Revolution.

PCK: In terms of subsidies to farmers under globalisation, the Indian farmers appear to have been let down. Your suggestion of a 'livelihood security box' during the early stages of negotiations under agriculture for trade under globalisation should have been included. In the meantime, the farmers' suicides are mounting, and rural youth would want to quit farming. The kind of agrarian crises and farmers' distress witnessed today might not be solved just by increasing the budgetary allocation to agriculture. As you have been repeatedly emphasising, profitability and sense of joy with pride in farming need to be restored. How could these be achieved? In fact, I would like to ask you what could make the best of the brains get involved in agriculture, animal husbandry and fisheries. Would it be just money or else money plus dignity, status in society etc?

MSS: During 1990–94 I pleaded for the inclusion of a 'livelihood security box' in the World Trade Agreement. The industrialised countries have got provisions like 'green box' and 'amber box' included in the world trade agreement. Unfortunately we lost a great opportunity for getting a livelihood security box which will help to protect the livelihood of farm families. Industrialised countries do not recognise that agriculture in our country is just not a commercial enterprise but is a very foundation of livelihood security. Agriculture has to undergo another technological revolution as in 1960. Such a technological revolution should be based upon the principles of ecology, economics, gender and social equity, renewable energy and climate management. This requires a systems approach to technology development and dissemination. The younger generation will be attracted to farming only if it is technologically advanced and is able to provide reasonable income. Ecology, economics and equity should be the bottom line. The younger generation needs opportunities for jobs with secure income. Modern industry promotes jobless growth, while agriculture leads to job led growth. Therefore the younger generation will have to get familiar with the uncommon opportunities available in crop and animal husbandry, inland and marine fisheries and agroforestry. Technologically qualified young people need intellectual satisfaction and stimulation, and not just money alone. A minimum essential income is however important.

PCK: What you have said above seems to suggest that there is a lot to be done. And much of it needs to be done by the government in the centre and states. The other thing is about the 'potential yield' and 'production gap' that we have been concerned about for quite some time. Can you please elaborate why the 'potential yield' still remains elusive?

MSS: The gap between potential and actual yield can be bridged only through an integrated package of technology, services and public policies. Public policies relating to input-output pricing and import-export decisions will play a very key role in making agriculture sustainable and in taking advantage of untapped potential.

PCK: The tendency is to talk about population stabilisation in India by about 2050 at around 1.50 to 1.60 billion. Some believe that the present population of about 1.30 billion is already far in excess of the 'carrying capacity' of India's Earth resources. I recall that the Working Group for defining the research and teaching programmes of the School of Life Sciences in late 1960s, of which you were the Chairman, had rightly included 'reproductive biology and contraception' as one among the other areas such as photobiology, radiation biology etc. Would it be a good idea to introduce incentives to families with 'zero' to 'one child'? Adopted children should not be included in the count. Further, it is also felt that the phrase 'Family Welfare' does not convey the urgent need for 'Family Planning'. What are your views on this? Population growth beyond biocapacity of the resources is a major obstacle in achieving the goal of a 'zero hunger' world.

MSS: Marquis de Condorcet mentioned that population will stabilise itself if children are born for happiness and not for just existence. Children by choice and not by chance should be our motto. For achieving this goal, a Population Committee I chaired in 1994 recommended the development of socio-demographic charters at the local level. In this context, it will be useful to call the Ministry of Family Planning rather than family welfare since it emphasises the need for planning for children's happiness. The local level socio-demographic charter will also help to stop current trend in sex ratios and will address the problems of child nutrition during the first 1000 days in a child's life.

PCK: Based on my review of major requirements for achieving food and nutrition security at the individual household level of millions of rural women and men, I find that you have been the first scientist to think of 'access to food' as important as the 'availability of food'. In the Princess Leelavathi Memorial Lecture which you delivered in 1972 in the University of Mysore, you dealt with the problem of rural poverty and also suggested ways to combat it. Your ideas of 'techniracy' with which the largely illiterate and unskilled rural women and men could be given skill empowerment, 'do ecology' etc., have been most innovative and relevant. While setting up the programmes of the M.S. Swaminathan Research Foundation, you have brought together ecoagriculture, ecotechnologies — driven on farm and non-farm ecoenterprises with market linkages for livelihoods as also satellite and computer based village knowledge centres for knowledge empowerment of young rural women and men. The other thing is the need to reduce the annual population explosion because the resources are rapidly dwindling but the consumers are quite rapidly increasing. It would help if you can kindly elaborate your advice and what the national policies ought to be.

MSS: I agree that poverty is the main cause of hunger and deprivation. Poverty can be overcome only through opportunities for meaningful employment. We should measure agricultural progress not only by the millions of tonnes of food produced but also by the progress in improving the net income of farmers. An income orientation involving agro-processing and value addition to primary products is essential.

About your other questions, you have rightly referred to the Princess Leelavathi Memorial Lecture that I was invited to deliver in 1972 at the University of Mysore. As you said, I mainly focused on rural livelihoods and how 'techniracy', a term coined by me to describe a 'pedagogic method of learning by doing', would help in the skill empowerment of the resource-poor, largely unskilled rural women and men. What we have to point out is the urgent need of 'Do Ecology'. A 'Don't Do' approach alone will not help since we have serious problems facing us like overcoming poverty, hunger and malnutrition. 'Do Ecology' will involve developing ecotechnologies to meet different environmental challenges like shrinking land resource, depleting ground water table etc.

The management of the monsoon behaviour and the challenge of conservation of natural resources will need appropriate biodiversity.

I have already mentioned that 'anthropocene' has to be viewed in a holistic way and it requires a systems approach to the management of natural resources. This applies also to the smallholder farms practising one or another kind of ecoagriculture. You know that I have been emphasising the need for every farm to have the following three components: (i) Nitrogen-fixing trees to minimise the application of inorganic nitrogen fertilisers, the production of which by chemical processes needs the burning of huge quantities of fossil fuels, and further it leads to accumulation of nitrates in the soil and aquifers; (ii) Biogas plant that provides fuel to cook and lighting in the rural houses, and at the same time effectively eliminates the release of greenhouse gases into atmosphere; (iii) A rainwater harvesting pond. If these are met, the small and medium farmers, in particular, can enjoy nutrition security and energy security.

I have also been pointing out that the production of fodder and forage needs to be greatly enhanced as India has a huge cattle population. The small and marginal farms with farm animals can use dung and urine not only for soil fertility enrichment, but also to produce bio-gas, and in turn, can provide fodder to them.

I must add that rain water harvesting is also very important for individual houses. In fact, rain water harvesting should become mandatory both in the farm and in the house so that domestic as well as farm needs can be met on a sustainable basis.

PCK: Thank you very much for sparing your valuable time. Despite the fact that humanity is currently facing many environmental and social challenges, you have shown the path to 'Biohappiness'.

Thank you for the wonderful time and highly informative conversation.

Bibliography

1. Dil, A. (2004). *Life and Work of M.S. Swaminathan. Toward a Hunger-free World*, East West Books (Madras) Pvt. Ltd., Chennai, p. 636.
2. Ramanujam, S., Siddiq, E.A., Chopra, V.L. and Sinha, S.K. (1980). *Science and Agriculture: M.S. Swaminathan and the Movement for Self-Reliance*, IARI, New Delhi, p. 400.
3. Erdelyi, A. (2002). *The Man Who Harvests Sunshine — The Modern Gandhi: M.S. Swaminathan*, Tertia kiadó, Budapest, p. 167.
4. Iyer, R.D. (2002). *Scientist and Humanist. M.S. Swaminathan*, Bharatiya Vidya Bhavan, New Delhi, p. 245.
5. Gopalakrishnan, G. (2002). *M.S. Swaminathan: One Man's Quest for a Hunger-Free World*, Education Development Centre, Chennai, p. 130.
6. Parasuraman, N. (2014). *M.S. Swaminathan–Architect of Sustainable Agriculture*, Mathi Nilayam, Chennai, p. 239 (in Tamil).
7. Deolgavkar, A. (2000). *Swaminathan Bhukmukticha Dhyas* (Liberator from hunger), Akshar Prakashan, Mumbai, p. 162.
8. Deolgavekar, A. (2008). *Swaminathan Bhukmukticha Dhyas*, Supriya Sharad Marathe, Mumbai, p.183 (in Marathi).
9. Swaminathan, M.S. (2010). *From Green to Evergreen Revolution,* Academic Foundation, New Delhi, p. 400.
10. Kesavan, P.C. and Iyer, R.D. (2014). M.S. Swaminathan: A journey from the frontiers of life sciences to a state of 'Zero Hunger World'. *Current Science* **107**, 2036–2051.
11. Kesavan, P.C. and Hariharan, G.N. (2015). Special section: Sustainable food and nutrition security. *Current Science* **109**(5), 415–512.
12. Fresco, L.O. (2015). The New Green Revolution: Bridging the gap between science and society. *Current Science* **109**(5), 430–438.

13. Wilson, E.O. (2003). *The Future of Life*, Vintage Books, London.
14. Rabbinge, R. (2015). M.S. Swaminathan: His contributions to science and public policy. *Current Science* **109**(5), 439–446.
15. Martin Rees (2003). *Our Final Hour: A Scientist's Warning*, Basic Books, New York.
16. Salaman, R.N. (1949). *The History and Social Influence of the Potato*, Cambridge Press, United Kingdom, p. 685.
17. Swaminathan, M.S. (1950). *Der Zuchter* **20**, 358–360.
18. Eigsti, O.J. and Dustin, P. (1955). *Colchicine in Agriculture, Medicine, Biology and Chemistry*, Iowa State College Press, Iowa, p. 496.
19. Mehta, R.K., Subramanyan, K.N. and Swaminathan, M.S. (1965). *Indian Journal of Genetics and Plant Breeding* **25**, 305–316.
20. Riley, R. and Chapman, V. (1961). *Journal of Heredity* **52**, 22–26.
21. Lukaszewski, A.J., Apolinarska, B., Gustafson, P. and Krolon, K.D. (1987). *Genome* **29**(4), 562–569.
22. Swaminathan, M.S. (1965). *Indian Journal of Genetics* **26A**, 29–41.
23. Swaminathan, M.S. (1969). Mutation Breeding. *Proceedings of the XII International Congress of Genetics* **3**, 327–347.
24. Prabhakar Rao, M.V. and Swaminathan, M.S. (1963). *Indian Journal of Genetics* **23**(3), 232–240.
25. Swaminathan, M.S. and Howard, H.W. (1953). *Bibliographica Genetica* XVI, 1–192.
26. Levitsky, G.A. and Benetzkaja, G.K. (1927). *Bulletin of Applied Botany, Genetics and Plant Breeding* **17**, 289–304.
27. Prakken, R. and Swaminathan, M.S. (1951). *Mededelingen van de Landbouwhogeschool Te Wageningen/Nederland, Decl 50, Verhandeling* **8**, 3–7.
28. Swaminathan, M.S. (1951). *The American Potato Journal* **28**, 472–489.
29. Swaminathan, M.S. (1955). *Nature* (Lond.) **176**, 887–888.
30. Prakken, R. and Swaminathan, M.S. (1952). *Genetica* **26**, 77–101.
31. Stebbins, G.L. (1947). *Advances in Genetics* **1**, 403–429.
32. Howard, H.W. and Swaminathan, M.S. (1952). *Euphytica* **1**, 20–28.
33. Swaminathan, M.S. (1954). *Genetics* **39**(1), 59–76.
34. Swaminathan, M.S. (1956). *Nature* (Lond.) **178**, 599–600.
35. Swaminathan, M.S. and Hougas, R.W. (1954). *American Journal of Botany* **41**(8), 645–651.

36. Swaminathan, M.S. (1954). *Journal of Heredity* **45**(6), 265–272.
37. Swaminathan, M.S. (1951). *American Potato Journal* **28**(1), 472–489.
38. Swaminathan, M.S. (1952). *New Biology* **13**, 31–48.
39. Swaminathan, M.S. (1958). *Indian Journal of Genetics* **18**, 8–15 (Special Symposium Number).
40. Wipf, L. and Cooper, D.C. (1938). *Proceedings of the National Academy of Sciences,* USA, **24**, 87–91.
41. Bhaskaran, S. and Swaminathan, M.S. (1958). *The Nucleus* 75–78.
42. Sikka, S.M., Mehta, R.K. and Swaminathan, M.S. (1958). *Nature* **181**, 32–33.
43. Howard, A. and Pelc, S.R. (1953). *Heredity* **6**, 261–273.
44. Bhaskaran, S. and Swaminathan, M.S. (1959). *Current Science* **28**, 335–336.
45. Kesavan, P.C. (2014). *Current Science* **107**(1), 46–53.
46. Muller, H.J. (1927). Artificial transmutation of the gene. *Science* **66**, 84–87.
47. Stadler, L.J. (1928). *Science* **68**, 186–187.
48. Gustafsson, A. (1947). *Hereditas* **33**, 1–100.
49. Sikka, S.M. and Swaminathan, M.S. (1955). *Euphytica* **4**, 173–182.
50. Swaminathan, M.S. and Natarajan, A.T. (1956). *Current Science* **25**, 279–281.
51. Swaminathan, M.S. and Natarajan, A.T. (1956). *Information Service* **4**, 5–7.
52. Swaminathan, M.S. and Natarajan, A.T. (1959). *Science* **130**, 1407–1409.
53. Muller, H.J. (1954). *Radiation Biology* **1**, 475–479.
54. Wolff, S. (1998). *Environ. Health Perspectives* **106**, 277–283.
55. Farooqui, Z. and Kesavan, P.C. (1993). *Mutation Research* **302**, 83–89.
56. Swaminathan, M.S. (1957). *Indian Journal of Genetics and Plant Breeding* **17**, 296–304.
57. Swaminathan, M.S. and Natarajan, A.T. (1957). *Nature* (Lond.) **179**, 479–480.
58. Swaminathan, M.S. (1958). *The Statesman* 6 October, 1958.
59. Bhaskaran, S. and Swaminathan, M.S. (1960). *Genetica* **31**, 449–480.
60. Bhaskaran, S. and Swaminathan, M.S. (1961). *Genetica* **32**, 1–32.
61. Kesavan, P.C. (2005). *Current Science* **89**(2), 318–328.
62. Natarajan, A.T., Sikka, S.M. and Swaminathan, M.S. (1958). *Proceedings of II V.N. International Conference on Peaceful Uses of Atomic Energy,* Geneva, **27**, 321–331.

63. Jagathesan, D., Bhatia, C.R. and Swaminathan, M.S. (1961). *Nature* (Lond.) **190,** 468.
64. Vervelde, G.J. (1953). *Netherlands Journal of Agricultural Science* **1,** 2–4.
65. Jagathesan, D. and Swaminathan, M.S. (1961). *Die Naturwissenschaften* **9,** 384–385.
66. Pai, R.A. and Swaminathan, M.S. (1960). *Evolution* **14**(4), 427–432.
67. Bhaskaran, S. and Swaminathan, M.S. (1960). *Experimental Cell Research* **20,** 598–599.
68. Pai, R.A., Upadhya, M.D., Bhaskaran, S. and Swaminathan, M.S. (1961). *Chromosoma (Berl.)* **12,** 398–409.
69. Nirula, S., Bhaskaran, S. and Swaminathan, M.S. (1961). *Experimental Cell Research* **24,** 160–162.
70. Swaminathan, M.S. (1963). *Comitato Nazionale Energia Nucleare, Estratto da: L´energia Nucleare in Agricoltura*, 243–277.
71. Darwin, C. (1859). *The Origin of Species by Natural Selection*, John Murray, London.
72. Swaminathan, M.S. and Prabhakara Rao, M.V. (1961). *Wheat Information Service* **13,** 9–10.
73. Bhatia, C.R. and Swaminathan, M.S. (1962). *Z. Pflanzenzuchtg* **48,** 317–326.
74. Gregory, W.C. (1956). *Brookhaven Symposia in Biology* **9,** 177–190.
75. Swaminathan, M.S. (1965). *Journal of the IARI Post-graduate School* **3,** 83–88.
76. Swaminathan, M.S. (1966). *Indian Journal of Genetics* **26A** (Symposia number), 29–41.
77. Swaminathan, M.S. (1962). *Current Science* **31,** 308–309.
78. Errera, M. and Forssberg, H. (1961). *Mechanisms in Radiobiology,* Volume I. General Principles, Academic Press, New York and London, p. 534.
79. Thoday, J.M. and Read, J. (1947). *Nature (Lond)* **160,** 608.
80. Gopal Ayengar, A.R. and Swaminathan, M.S. (1966). Use of neutron irradiation in agriculture and applied genetics. *Proceedings of IAEA Symposium on Biological Effects of Neutron and Proton Irradiations.* Volume I, Brookharen National Lab, New York, pp. 409–432.
81. Stone, W.S., Wyss, O. and Haas, F. (1947). *Proceedings of National Academy of Sciences, USA* **33,** 59–66.
82. Stone, W.S. (1955). *Brookhaven Symposia on Biology* **8,** 171–190.

83. Natarajan, A.T. and Swaminathan, M.S. (1958). *Indian Journal of Genetics Plant Breeding* **18**, 220–223.
84. Swaminathan, M.S., Nirula, S., Natarajan, A.T. and Sharma, R.P. (1963). *Science* **141**(3581), 637–638.
85. Chopra, V.L., Natarajan, A.T. and Swaminathan, M.S. (1963). *Radiation Botany* **3**, 1–6.
86. Chopra, V.L., Natarajan, A.T. and Swaminathan, M.S. (1963). *Die Naturwissenschaften* **10**, 374–375.
87. Nirula, S., Sharma, R.P., Swaminathan, M.S. and Natarajan, A.T. (1963). *Drosophila Inform. Serv.* **38**, 1.
88. Swaminathan, M.S., Chopra, V.L. and Bhaskaran, S. (1962). *Radiation Research* **16**(2), 182–188
89. Kesavan, P.C. and Swaminathan, M.S. (1967). *Radiation Botany* **7**, 269–272.
90. Kesavan, P.C. and Swaminathan, M.S. (1970). *Radiation Botany* **10**, 199–205.
91. Kesavan, P.C. and Swaminathan, M.S. (1969). *Indian Journal of Genetics* **29**(2), 173–183.
92. Kesavan, P.C. and Swaminathan, M.S. (1966). *Current Science* **35**, 403–404.
93. Kesavan, P.C. and Swaminathan, M.S. (1971). *Radiation Botany* **11**, 253–281.
94. Ganesan, A.T. and Swaminathan, M.S. (1958). *Stain Technology* (USA) **33**, 115–121.
95. Swaminathan, M.S. and Ganesan, A.T. (1958). *Nature* **182**, 610–611.
96. Swaminathan, M.S. and Ganesan, A.T. (1958). *Memoirs of Indian Botanical Society* **1**, 111–117.
97. Brewbaker, J.L. and Swaminathan, M.S. (1960). *Current Science* **29**, 298–301.
98. Bauer, R. (1957). *Hereditas* **43**, 323–329.
99. Swaminathan, M.S. (1961). Proceedings of the IAEA (International Atomic Emergy Agency), *Symposium on Effects of Ionizing Radiation on Seeds, and their significance for Crop Improvement, Karlsruhe*, pp. 279–288.
100. Swaminathan, M.S. and Natarajan, A.T. (1956). *Current Science* **25**, 382–385.
101. Swaminathan, M.S. and Natarajan, A.T. (1957). *Stain Technology* **32**(1), 43–45.
102. Swaminathan, M.S. and Natarajan, A.T. (1959). *Journal of Heredity* **50**, 177–87.
103. Chopra, V.L. and Swaminathan, M.S. (1966). *Indian Journal of Genetics* **26**, 59–62.

104. Kumar, S., Aggarwal, U. and Swaminathan, M.S. (1967). *Mutation Research* **4**, 155–162.

105. Prasad, M.V.R., Krishnaswamy, R. and Swaminathan, M.S. (1967). *Current Science* **36**(16), 438–439.

106. Savin, V.N., Swaminathan, M.S. and Sharma, B. (1968). *Mutation Research* **6**, 101–107.

107. Varughese, G. and Swaminathan, M.S. (1968). *Indian Journal of Genetics* **28**(2), 158–165.

108. Swaminathan, M.S., Siddique, E.A., Singh, C.B. and Pai, R.A. (1970). In: Rice Breeding with Induced Mutations II (Technical Reports Series No. 102), IAEA, Vienna, pp. 25–43.

109. Siddique, E.A. and Swaminathan, M.S. (1968). *Mutation Research* **6**, 478–481.

110. Swaminathan, M.S. (1969). *Proceedings of the XII International Congress of Genetics* **3**, 327–347.

111. Swaminathan, M.S. (1969). Short Review of the IAEA Symposium held at Washington State University, Pullman, Washington, USA, July 1969, pp. 735–736.

112. Wallace, A.T. (1965). *Radiation Botany Supplement* **5**, 237–250.

113. Upadhya, M.D. and Swaminathan, M.S. (1969). *Indian Journal of Genetics* **29**, 338–341.

114. Bhaskaran, S. and Swaminathan, M.S. (1962). *Radiation Botany* **1**, 166–181.

115. Swaminathan, M.S. and Upadhya, M.D. (1964). *Current Science* **33**(15), 472–473.

116. Subramanian, M.K. and Subramanyan, S. (1961). *Current Science* **30**, 172.

117. Subramanian, M.K. and Subramanyan, S. (1964). *Current Science* **33**, 217.

118. Taylor, J.H. (1963). *Molecular Genetics*, Part I, Academic Press, New York, pp. 65–111.

119. Bastia, D. and Swaminathan, M.S. (1967). *Experimental Cell Research* **48**, 18–26.

120. Gall, H. (1963). *Science* **139**, 120.

121. DuPraw, E.J. (1965). *Proceedings of the National Academy of Sciences* (Washington), **53**, 161–180.

122. Swaminathan, M.S. (1968). The age of algeny, genetic destruction of yield barriers, and agricultural transformation. *Presidential Address, Section of*

Agricultural Sciences: 55ᵗʰ Indian Science Congress, Part II, Varanasi, pp. 236–248.

123. Swaminathan, M.S. (1984). Genetic Conservation: Microbes to Man. *Presidential Address, XV International Congress of Genetics.* In: Genetics, New Frontiers. Chopra, V.L., Joshi, B.C., Sharma, R.P. and Bansal, H.C. (Eds.) Vol. 1, Oxford and IBH Publishing Co., New Delhi, pp. 29–56.

124. Ehrlich, P. (1968). *The Population Bomb,* Sierra Club/Ballentine Books, USA, p. 201.

125. Paddock, W. and Paddock, P. (1967). *Famine 1975! America's Decision: Who Will Survive?,* Little Brown and Co., USA, p. 286.

126. Swaminathan, M.S. (Ed.) (1963). *Wheat Revolution: A Dialogue,* MacMillan India Ltd., Madras, p. 164.

127. Vogel, O.A., Craddock, J.C., Muir, C.E., Everson, E.H. and Rhode, C.H. (1956). *Agronomy Journal* **48**, 76–78.

128. Kumar, S., Bansal, H.C., Singh, D. and Swaminathan, M.S. (1967). *Z. Pflanzenzuchtung* **57**(4), 317–324.

129. Carson, R. (1962). *The Silent Spring,* Houghton Miffin Co., Boston, p. 400.

130. Swaminathan, M.S. (1965). *Illustrated Weekly of India* May 11, 1969.

131. Stevenson, J.R., Villoria, N., Byerlee, D., Kelley, T. and Maredia, M. (2013). *Proceedings of the National Academy of Sciences,* USA. **110**, 8363–8368.

132. Swaminathan, M.S. (2002). *From Rio de Janeiro to Johannesburg: Action Today and Not just Promises for Tomorrow,* Eastwest Books Pvt. Ltd., Chennai, p. 224.

133. Bourne Jr, J.K. (2009). *National Geographic* **215**(6), 26–59.

134. Dhillon, B.S., Kataria, P. and Dhillon, P.K. (2010). *Current Science* **98**(1), 33–36.

135. DeFries Ruth (2014). *The Big Ratchet: How Humanity Thrives in the Face of Natural Crisis,* Basic Books, New York, p. 273.

136. Kesavan, P.C. (2015). *Current Science* **108**, 1550–1551.

137. Wilson, E.O. (2002). *The Future of Life.* Vintage Books, London.

138. Swaminathan, M.S. (1972). *Agricultural Evolution, Productive Employment and Rural Prosperity,* Princess Leelavathi Memorial Lecture. January 17, 1972, University of Mysore, p. 35.

139. Swaminathan, M.S. (1973). *Everyday Science* **XVIII**, 1–6.

140. Brundtland, G.H. (1987). *Our Common Future. World Commission on Environment and Development*, Oxford University Press, Oxford, p. 416.

141. Swaminathan, M.S. (1987). *Food 2000: Global Policies for Sustainable Agriculture.* World Commission on Environment and Development, Zed Books Ltd., London and New Jersey, p. 131.

142. Swaminathan, M.S. (1996). *Sustainable Agriculture: Towards an Evergreen Revolution*, Konark Publishers Pvt. Ltd., Delhi, p. 232.

143. Swaminathan, M.S. (1996). *Sustainable Agriculture: Towards Food Security*, Konark Publishers Pvt. Ltd., Delhi, p. 272.

144. Swaminathan, M.S. (1999). *I Predict: A Century of Hope. Towards an Era of Harmony with Nature and Freedom from Hunger*, Eastwest Books Pvt. Ltd., Madras, p.155.

145. Swaminathan, M.S. (2000). An evergreen revolution. *Biologist* **47**(21), 85–89.

146. Swaminathan, M.S. (1998). Environmental Protection and Livelihood Security of the Rural Poor. Second Indira Priyadarshini Memorial Lecture, 19th November 1989, Krishak Bharati Co-operative Ltd., New Delhi, p. 10.

147. Rockström, J., Steffen, W., Noone, K., Persson, A., Chapin, F.S., III, Lambin, E.F., Lenton, T.M., Scheffer, M., Folke, C., Schellnhuber, H.J., Nykvist, B., de Wit, C.A., Hughes, T., van der Leeuw, S., Rodhe, H., Sörlin, S., Snyder, P.K., Costanza, R., Svedin, U., Falkenmark, M., Karlberg, L., Corell, R.W., Fabry, V.J., Hansen, J., Walker, B., Liverman, D., Richardson, K., Crutzen, P. and Foley, J.A. (2009). A safe operating space for humanity. *Nature* **461**, 472–475.

148. Swaminathan, M.S. and Kesavan, P.C. (2016). *Current Science* **110**, 7–9.

149. Swaminathan, M.S. (1973). Agriculture on spaceship earth. Third coromandel lecture, *Fertilizer News* **18**(4), 71–86.

150. Goldsmith, E. and Prescott–Allen, R. (1972). *A Blueprint for Survival — The Ecologist.* Penguin Books, London.

151. Kesavan, P.C. and Swaminathan, M.S. (2012). *Evergreen Revolution in Agriculture — Pathway to a Green Economy*, Westville Publishing House, New Delhi, p. 139.

152. Kesavan, P.C. and Swaminathan, M.S. (2007). Future direction of Asian agriculture: Sustaining productivity without ecological degradation. *ASM Science Journal* **1**, 161–168.

153. Kesavan, P.C. and Swaminathan, M.S. (2008). Strategies and models for agricultural sustainability in developing Asian countries. *Philosophical Transactions of the Royal Society B.* **363**, 877–891.

154. Kesavan, P.C. and Swaminathan, M.S. (2006). From green revolution to evergreen revolution: Pathways and terminologies. *Current Science* **90**(2), 145–146.

155. McNeely, J.A. and Scherr, S.J. (2003). *Ecoagriculture: Strategies to Feed the World and Save the Wild Biodiversity,* Island Press, Washington D.C., p. 327.

156. Higa, T. (1996). *An Earth Saving Revolution: A Means to Resolve our World's Problems through Effective Microorganisms (EM),* Sunmark Publishing Co., Tokyo, p. 336.

157. Fukuoka, M. (2009). *The One-Straw Revolution.* Original copy right 1978; Introduction Copyright 2009 by Frances Moore Lappé. The New York Review of Books, ISBN 978-1-59017-313-8.

158. Kesavan, P.C. and Malarvannan, S. (2010). *Current Science* **99**(7), 908–914.

159. Swaminathan, M.S. (2014). Zero Hunger (editorial) *Science* **345**, 461.

160. Kesavan, P.C. and Swaminathan, M.S. (2014). *Current Science* **107**(12), 1970–1974.

161. Pretty, J. (2008). *Philosophical Transactions of the Royal Society B (London)* **363**, 447–465.

162. Uma Lele and Kavitha Gandhi (2009). *M.S. Swaminathan Research Foundation at '21.* Report of the Independent Program Review, Chennai, p. 313.

163. Meadows, D.H., Meadows, D.L., Randers, J. and Behrens, W.W. (1972). *The Limits to Growth: A Report for the Club of Rome's Project on the Predicament of Mankind,* Universe Books, New York, p. 211.

164. Meadows, D., Randers, J. and Meadows, D. (2004). *Limits to Growth — The 30-Year Update,* Chelsea Green Publishing Company, New York, p. 338.

165. Nitya Rao (2015). *M.S. Swaminathan in Conversation with Nitya Rao: From Reflections On My Life to the Ethics and Politics of Science,* Academic Publisher, New Delhi, p. 227.

166. Wakernaagel, M., Schulz, N., Deumling, D., Linares, A., Jenkins, M., Kapos, V., Monfreda, C., Loh, J., Myers, N., Norgaard, R. and Randers, J. (2002). Tracking the ecological overshoot of the human economy. *Proceedings of the National Academy of Sciences* **99**(14), 9266–9271.

167. Swaminathan, M.S. (Ed.). (1994). *Ecotechnology and Rural Development: A Dialogue.* MacMillan India Ltd., Madras, p. 396.

168. Swaminathan, M.S. (Ed.). (1993). *Information Technology: A Dialogue.* MacMillan India Ltd., Madras, p. 264.

169. Swaminathan, M.S. (2015). *Combating Hunger and Achieving Food Security,* Cambridge University Press, New Delhi, p. 167.

170. Swaminathan, M.S. (2012). *Remember Your Humanity: Pathway to Sustainable Food Security.* India Publishing Agency, New Delhi.

171. Swaminathan, M.S. (2010). *Science and Sustainable Food Security: Selected Papers of Swaminathan, IISc Centenary Lecture Series.* IISc Press, World Scientific Publishing Co. Pte. Ltd., Singapore, p. 420.

172. Swaminathan, M.S. (2009). Gene banks for a warming planet. *Science* **325**, 517.

173. Swaminathan, M.S. (2015). *In Search of Biohappiness,* 2nd edition, World Scientific Publishing Co. Pte. Ltd., Singapore, p. 205.

174. Steffen, W., Persson, A., Deutsch, L., Zalasiewicz, J., Williams, M., Richardson, K., Crumley, C., Crutzen, P., Folke, C., Gordon, L., Molina, M., Ramanathan, V., Rockström, J., Scheffer, M., Schellnhuber, H.J., and Svedin, U. (2011). The anthropocene: From global change to planetary stewardship, *Ambio* **40**(7), 739–761.

175. Swaminathan, M.S. (1973). *Our Agricultural Future,* Sardar Patel Memorial Lectures, All India Radio, p. 54.

176. Swaminathan, M.S. (1998). El Nino and monsoon management. *Environmental Awareness* **21**(2), 46–48.

177. Swaminathan, M.S. (2005). Beyond Tsunami: An agenda for action. *The Hindu* 17 January, 2005.

178. Sinha, S.K. and Swaminathan, M.S. (1991). *Climate Change* **19**, 201–209.

179. Nature Editorial (2011). The mask slips — The Durban meeting shows that climate policy and climate science inhabit parallel worlds. *Nature* **480**, 292.

180. Swaminathan, M.S. and Kesavan, P.C. (2012). Agricultural Research in an era of climate change. *Agricultural Research,* **1**(1), 3–11.

181. Swaminathan, M.S. (2015). Achieving the goal of food security and nutrition security for all. *NFI Bulletin* **36**, 1–5.

182. Swaminathan, M.S. (1986). Building national and global nutrition security systems. In: *Global Aspects of Food Production.* Natural Resources and the

Environment Series. Swaminathan, M.S. and Sinha, S.K. (Eds.) Tycooly International Press, Oxford–Riverton, Dehradun, **20**, pp. 417–449.

183. Swaminathan, M.S. (1991). Agriculture and food systems. In: *Climate Change: Science, Impacts and Policy.* Proc. 2nd World Climate Conf. J. Jager and H.L. Ferguson (Eds.) Cambridge University Press, Cambridge, pp. 269–277.

184. Swaminathan, M.S. (2012). Combating Hunger. *Science* **338**, 1005.

185. Swaminathan, M.S. (2014). Zero Hunger. *Science* **345**, 461.

186. Das, P., Bhavani, R.V. and Swaminathan, M.S. (2014). A farming system model to leverage agriculture for nutritional outcomes. *Agricultural Research* **3**(3), 193–203.

Index